The Structure Of The Alps

TO MY FRIEND
E. B. BAILEY, F.R.S.

Léon W. Collet

(1880 – 1957)

(photograph courtesy of Mrs. L.W. Collet)

The Structure Of The Alps

By
LÉON WILLIAM COLLET, 1880 D. Sc., LL. D.
Professor of Geology and Paleontology in the
University of Geneva; Sometime Professor of
Alpine Geology in The Harvard University

Foreward by
O. T. JONES, D. Sc., F. R. S.
Woodwardian Professor of Geology in the
University of Cambridge

With Introduction by
ALBERT V. CAROZZI
Professor of Geology,
University of Illinois
at Urbana-Champaign, Illinois

ROBERT E. KRIEGER PUBLISHING COMPANY
HUNTINGTON, NEW YORK
1974

Original Edition 1927
Second Edition 1935, Reprint 1974

Printed and Published by
ROBERT E. KRIEGER PUBLISHING CO. INC.
645 NEW YORK AVENUE
HUNTINGTON, NEW YORK 11743

© Copyright 1974 — introduction by
Robert E. Krieger Publishing Co. Inc.

Library of Congress Cataloging in Publication Data

Collet, Léon William, 1880-1957.
    The structure of the Alps.

    Reprint of the 1935 ed. published by E. Arnold, London.
    Includes bibliographies.
    1. Geology--Alps.   I. Title.
QE285.C67 1974   554.947   73-92864
ISBN 0-88275-150-6

Printed in U.S.A. by
NOBLE OFFSET PRINTERS, INC.
New York, N.Y. 10003

FOREWORD

BY PROFESSOR O. T. JONES, D.Sc., F.R.S.

This account of the Structure of the Alps, in which is brought together the results of the investigations of a numerous band of distinguished geologists, will prove a great boon to the ever-increasing number of English-speaking geologists and geographers who take an active interest in the geological history and tectonic problems of that fascinating region. Professor Collet, by accurate and painstaking work in the field, has himself played an important part in elucidating some of these problems. He is well known, both to geographers and geologists in this country, as an enthusiastic and charming exponent of Alpine tectonics, and there are many besides who have been privileged to take part in the delightful excursions which he has conducted from time to time to some of the most interesting regions described in the book.

I retain a vivid recollection of the pleasant time that Professor W. J. Pugh and I spent in the company of Professor Collet and his collaborator, Dr. Paréjas, while they were mapping the Jungfrau mass. We had impressed upon us in a very practical manner the arduous nature of geological investigations in the high Alps, and the demands it makes upon the physical powers of the investigator. It cannot be too strongly impressed upon readers of the story of the Alps as unfolded in this book, that the results obtained by the brilliant band of investigators to whom Professor Collet makes reference, as synthesized by Staub and elucidated by the genius of Argand, have only been acheived by heroic endeavours often in the face of unforeseen and unforeseeable dangers. Many of the clues to the tectonic puzzle lie high up in the face of almost inaccessible precipices, to reach which entails many hours of arduous climbing during which the investigator is exposed to the vagaries of the weather or the perils of snow avalanches, or the still greater perils of avalanches of stones. The fruits of these labours are garnered in Professor Collet's book. When it is borne in mind also that the season during which the high regions of the Alps are accessible is extremely short, one cannot but marvel at the progress which has been made in the last quarter of a century or so, in the elucidation of Alpine tectonics. There is no doubt that one of the results of having the geology of the Alps brought so tersely before English readers will be a still further widening of interest in the region. To those who are prompted to visit the Alps for the first time, or to those others who wish to renew their acquaintance with them, Professor Collet's book will equally form an invaluable guide and inspiration.

The extension of the scope of the book to include the Alpine Range in the Western Mediterranean will enlarge its usefulness and widen its appeal to these readers.

O. T. JONES.

PREFACE TO SECOND EDITION

As the first issue of this book has been out of print for a year, a second edition is necessary.

Since the publication of the first edition, in 1927, our knowledge of the structure of the Alps has been advancing. Further, a brilliant international gathering of geologists has been engaged in following the Alpine structure in the Western, as well as in the Eastern, Mediterranean regions, and new ideas of the structure of the Alpine Range have been presented.

Though this second edition embodies the general frame of the first, it has been revised and enlarged by a new part on the *Alpine Range in the Western Mediterranean*.

The relations between the Alps and Apennines are discussed, based on the window of the Alpi Apuane. The description of the tectonics of Corsica, Sardinia, the Island of Elba, the Southern chain of Spain, the Balearic Islands will enable the reader to follow the Alpine chain as far as Gibraltar and the Rif of Morocco.

Structures play such a great rôle in modern Geology and Physical Geography that students of Geology and Geography and general readers must be presented with the most up-to-date principles and the results of recent research. As shown in this second edition " Foundation Folding " is now a well-defined principle. The *Alpine Geosyncline*, since the study of the Apennines, appears now to be more complicated than it was thought ten years ago. Indeed, Kober's theory of the " Double thrusting " of the Alpine chain must be discussed, for his theory of " Betwixt-Mountains " seems to throw new light on the study of the Alpine geosynclinal.

The chapters on Mont Blanc and the Aiguilles Rouges and that on the Geology of the Jungfrau have been rewritten. The chapter on the Jura Mountains has been shortened to give more space to the Alps, following the suggestion of friends.

Several sections of the first edition have been replaced and new sections have been added. In order to avoid a higher price for the book, unnecessary details have been cut and the bibliography limited to important papers.

This second edition is based on various courses of lectures by the author in the University of Geneva and, during three half years, in the University of Harvard, in combination with personal research work in the field : in the Apennines, the Island of Elba, the Balearic Islands and Algeria, as well as the Alps.

The author thanks his friends for kind criticism of the first edition and useful suggestions for the second.

He also records his grateful appreciation to the help of his assistant, Mr. P. Vaugelas, for the drawing of the new figures, and of Mr. Arnold Lillie, one of his research students, for reading and revising the English of the manuscript, and for valuable suggestions from the point of view of an advanced student.

The best thanks of the author are also due to his friends, Professor E. B. Bailey and Professor Ed. Paréjas, for the critical reading of the book in typescript or in proof.

The author is greatly indebted to Mr. F. P. Dunn (Messrs. Edward Arnold & Co) whose co-operation has been materially responsible for the success of the first edition and for the illustrations of the second.

GEOLOGICAL DEPARTMENT,

   THE UNIVERSITY,

                                 LÉON W. COLLET.

   GENEVA, *October*, 1935.

PREFACE TO FIRST EDITION

Apart from a few of my contributions, there is no recent publication in the English language dealing with the Structure of the Alps. For the ordinary student and the general reader, books in foreign languages on the Structure or on the Geology of the Alps are difficult to read, for they are either too technical or too detailed. It was accordingly suggested to me that I should write an account of the subject which might be of use not only to students but also to the general reader.

It is obvious that the structure of every part of the Alps cannot be dealt with in this kind of book, and that certain regions had to be selected as typical examples. Since many British geologists, geographers and members of other learned professions spend their vacations among the Alps, those parts which they are most likely to see have been described in greater detail, e.g., The Salève (near Geneva), Mont-Blanc, Zermatt, Lauterbrunnen, Grindelwald, Bernina Pass, Maloggia Pass, Bregaglia valley and Aosta valley. Moreover, the reader will find detailed itineraries of the surroundings of Zermatt, Lauterbrunnen, Grindelwald and Geneva which will enable him to follow several of the arguments in the field.

Stratigraphical sequences have been quoted only when especially interesting, or in order to enable the reader to understand correlations; while it was necessary, in many cases, to deal with questions relating to paleogeography. The sections given in the text or in folding plates are not supposed to give stratigraphical details, but are primarily intended to show structures.

Owing to the limited size of the book, it was not possible to include maps showing all the names mentioned in the text. The reader should consult either a good topographical map of Switzerland or the geological maps mentioned in the bibliographies at the end of Chapters or Parts of the book.

I am much indebted to friends and colleagues who did their best to enable me to give good illustrations of the Swiss Alps.

Professor M. Lugeon very generously placed at my disposal the splendid, and as yet unpublished, photograph of the Dents de Morcles on which he marked the geological contours.

Professor Emile Argand gave me permission to reproduce the important plate on the Formation of the Western Alps and his very valuable plate on the Tectonics of the Pennine Alps, near Zermatt.

From Professor Albert Heim, Professor Arbenz, Professor Buxtorf and Dr. Staub I have received permission to reproduce interesting sections.

The figures have been culled from many different sources, but they have usually been simplified for this special purpose. I am greatly indebted to my chief assistant Dr. Ed. Paréjas for drawing these figures and especially for the reduction of R. Staub's geological map.

My best thanks are due to my friends E. B. Bailey, Dr. G. W. Lee and Professor S. J. Shand for the critical reading of the whole or part of the book in typescript or in proof.

L. W. C.

August, 1927.

CONTENTS

PART I

INTRODUCTION

CONTENTS

PART II

THE FORELAND

PART III

THE GEOSYNCLINE. THE WESTERN ALPS OR THE PENNINE NAPPES

CONTENTS

PART IV

BETWIXT-MOUNTAINS
THE EASTERN ALPS OR THE AUSTRIDES

CONTENTS

LIST OF PLATES

INTRODUCTION

by

Albert V. Carozzi

Professor of Geology, University of Illinois

at Urbana-Champaign, Illinois

When, as a freshman in the fall of 1943, I timidly walked into the office of Professor Léon W. Collet at the University of Geneva, I was actually trying to evade my father's decree to become an organic chemist in spite of my longtime interest in geology. Professor Collet cut short my explanations and, in his rude but benign manner, handed me a volume by the famous French geologist Pierre Termier, entitled "A la gloire de la terre: souvenirs d'un géologue" (Paris, Desclée de Brouwer, 1924) and said: "Read this and see me tomorrow." After a night spent reading the enthusiastic prose of Pierre Termier I reached my decision to become a geologist and my father complied.

In this way I entered the world of Professor Collet, a department that he ran with fairness and overwhelming enthusiasm, but as a tight ship. Although his teaching assistant, I remember tiptoing in the hall so that "the bear," as the students nicknamed him, would not come out roaring from his lair.

Professor Collet belongs to the generation of Swiss geologists who had a profound worship for the mountains and the Alps in particular. While on mountain climbing trips he would not tolerate anybody ahead of him, and upon reaching 4000 meters elevation he would pull out from his backpack a white hard collar and a tie, putting both on in a mystical ritual.

His teaching was brilliant and concise. He used to make his points by means of dramatic formulas that a student would never forget for his entire life such as the Salève mountain, which overlooks Geneva, being "a knee-like anticline." Professor Collet was a convinced follower of continental drift as expressed by Alfred Wegener and applied to the Alps by Emile Argand. When, on top of the High Calcareous Alps, he described to his students the features of the Prealps, he would startle them saying: "You are looking at portions of Africa thrusted over the southern margin of Europe." When supervising his shivering students drawing a panorama of lofty peaks, he would criticize their sketches as lacking character, and he would thunder that the Alps "ne sont pas de la tomme!" namely, do not consist of soft cheese. He was himself a great believer in drafting large geological sections of valleys on the spot, even after many hours of arduous climbing, thus following the technique of Horace Bénédict de Saussure. Collet's sketches and paintings were remarkable, stressing the essentials with great strokes in a manner similar to the landscapes of the famous Swiss painter Ferdinand Hodler.

Professor Collet was born September 23, 1880 at Fiez, canton of Vaud. After finishing highschool at the Collège de Genève, he entered the University of Geneva where he obtained in 1904 his doctorate in physical sciences with a dissertation on the geology of the Tour Saillère- Pic de Tanneverge range in the High Calcareous Alps, under the supervision of Professor Charles Sarasin. Immediately after completing his degree he went to Edinburgh as an assistant to Sir John Murray, director of the Challenger Office. This introduction to the newly opened field of physical oceanography filled him with enthusiasm and led to the publication of his remarkable volume "Les dépôts marins" (Paris, 1908, Octave Doin, 325 p.) which was awarded the Jules Girard Price of the Geographic Society of Paris. He also spent some time as an assistant with the Scottish Lake Survey and developed a keen interest in physical limnology. Since his first stay in the British Isles, Professor Collet made numerous friends and always felt at home in that country. Nevertheless, he would not forget his beloved mountains and in 1911 he published a memoir entitled "Hautes-Alpes Calcaires entre Arve et Rhône," a masterpiece of Alpine field geology, which received the Plantamour-Prévost Award of the University of Geneva.

In 1912, after a short stay at the University of La Plata in Argentina, he was appointed Director of the Swiss Federal Hydrographic Service, a position he occupied with great efficiency until 1918 when he became professor of geology at the University of Geneva, following the retirement of Professor Sarasin. A few years later, in 1925, his treatise entitled "Les Lacs" (Paris, Octave Doin, 320 p.) was published. This book, which combined his wide experience on Scottish and Swiss lakes, became a milestone in the field of limnology, not only for its purely scientific value but also because of its emphasis toward practical applications at a time when the construction of numerous dams in the Alpine valleys was precisely stressing such a need.

In order to please his numerous British friends, in particular his longtime companion Sir Edward B. Bailey of the Geological Survey of Great Britain, and at the same time to spread the knowledge of Alpine geology to the English-speaking scientific community, he undertook the writing of "The Structure of the Alps," which was published in 1927 and followed by a second enlarged edition in 1935.

This volume, dedicated to E.B. Bailey "in recollection of many sunny days in the Swiss Alps and of glorious days in Scotland," is the most striking example of Professor Collet's talent to present a difficult subject in a clear, simple and well-illustrated fashion. The success of the volume was immense and countless foreign geologists found in it the ideal introduction to the Alps as well as a guidebook for planning a fieldtrip in these spectacular mountains. Although since 1935 much progress has been accomplished in the understanding of the Alps, the few published syntheses are technical papers—in French or German—sparingly illustrated while Collet's volume still remains the most easily understandable and illuminating work on the subject.

The publication of "The Structure of the Alps" led Harvard University to invite its author as a professor of Alpine Geology during the winter semesters of 1927 to 1929, giving him the opportunity to investigate aspects of the structure of the Rocky Mountains.

After twenty-six years of teaching and of training many students, of whom I am privileged to have been one, Professor Collet retired in 1944 and was appointed emeritus professor.

During his retirement of thirteen years, he continued his research activities, supported by a great physical strength he had acquired through lifelong mountain climbing. He was for many years a member of the Swiss Geologic Commission (1925-1953) and President of the Swiss Geologic Society from 1944 to 1948. He died on October 13, 1957 as the age of seventy-seven.

When the long and painful disease that was going to be fatal struck him, he withdrew to his mansion on the hill of Cologny, near Geneva, yet always happy to welcome colleagues and former students for a cup of tea. Often, when I visited him, he would be painting Alpine peaks from sets of photographs and greet me with such remarks as, "you are coming to see how the old mountain goat is doing, eh?" I believe that it is this sense of humour combined with his inspiring enthusiasm for nature that makes "The Structure of the Alps" an exciting book on which time has been unable to make any inroads.

Urbana, Illinois, December 1973

The Structure Of The Alps

THE STRUCTURE OF THE ALPS

PART I

INTRODUCTION

CHAPTER I

THE SITUATION OF THE ALPS IN THE EUROPEAN MOUNTAIN RANGES OF ALPINE AGE

The Alps, like a gigantic bow, extend from the Ligurian Sea to Vienna. They are framed at their outer margin by massifs of Hercynian age, such as the *Central Plateau* of France to the west, the *Vosges* and the *Black Forest* to the north, and farther to the east, the *Böhmer Wald*.

Ranges radiate from the Alps. With the latter, they constitute the European mountain ranges of Alpine age. Stretching from Andalusia to Asia, they connect the Alps with the Himalayas.

When the eastern extremity of the Alps dies away in the Hungarian Plain, the *Carpathians* come off the Alps to the north and the *Dinaric Alps* to the south. The Hungarian Plain lies between the two ranges. The *Balkan Mountains* are the continuation of the Carpathians to the east and of the Caucasus to the west.

The Dinaric Alps trend to the south and pass into Asia Minor.

At the western extremity of the Alps the *Apennines* may be followed to Sicily and even to the Atlas. The ranges of Provence extend as far as the *Pyrenees*.

The *Jura Mountains*, being only a " virgation " (see page 12) of the Alps, have to be considered as belonging to them.

One may ask whether there is a continuation of the Alps towards the west ? According to Argand and Staub a prolongation of the Alps may be seen in Corsica, in the Balearic Islands, in the Betic Cordillera (South of Spain), at Gibraltar and in the Rif of Morocco.

CHAPTER II

THE IMPORTANCE OF THE STUDY OF THE ALPS

Being at once the greatest and the youngest relief of Europe, the Alps play a rôle of paramount importance in *modern Geology.*

Since they are largely composed of sedimentary rocks, which have been folded, forming great ranges and even mountains of more than 4,000 metres (Great Combin), we arrive at the result that the Alps were formed by sediments deposited in a sea basin, called a geosyncline.

Being the youngest mountain range of Europe, the Alps show deep valleys, due to river erosion, in which the geologist may study the abnormal sequence of the strata, produced by orogenic movements. In these high cliffs—from Lauterbrunnen up to the summit of the Jungfrau there is a rise of 3,000 metres—the geologist sees only bare rocks, and the rare cover of scree or snow is no hindrance compared to the vegetation such as masks much of the South-West Highlands of Scotland.

But if at present the foreign geologist can enjoy geological panoramas, like that seen from the Gornergrat, without difficulties or danger, it must be remembered that Swiss geologists had to climb up the cliffs and gullies, roped together, with hammer and ice axe in their hands, to determine the age of the rocks, to follow the mechanical contacts, in a word to map out the geology of the mountains. In this work climbing is not a goal in itself ; it only furnishes the ways and means to get to the goal.

Since it has been recognized that continents are made up of mountain ranges of different ages, more or less " peneplained," it is clear that the results arrived at during the geological exploration of the Alps represent a great advance for Geological Science. Indeed, when the geometry, viz., the relations, of the different folds forming the Alps has been ascertained, the straightening out of these folds permits us to reconstitute the paleogeography, at various stages. But this can only be done on the basis of detailed stratigraphical studies, including consideration of the different facies and knowledge of the formation of marine deposits.

The study of the Geology of the Alps is the study of a great geological synthesis, based on evidence that can be investigated in the field. Indeed, it is a fascinating scientific research, for one has to deal with magnificent scenery forming a part of our ever-changing Earth.

H. B. de Saussure, the founder of Alpine Geology, was right when he said, at the end of the eighteenth century, that the study of mountains would permit us, better than anything else, to arrive at a Theory of the Earth.

CHAPTER III

THE DIFFERENT PARTS OF THE ALPS

Western and Eastern Alps. Long ago the Alps were divided, from a geological point of view, into :

1. *The Western Alps.*
2. *The Eastern Alps.*

The Western Alps extend from the Ligurian Sea to a line connecting the Lake of Constance with the Septimer Pass. The Eastern Alps stretch from the latter boundary to Vienna.

The Subdivisions of the Western Alps. From north to south, the principal subdivisions of the Western Alps are as follows :

1. *The Jura Mountains.*
2. *The Swiss Plateau*, made up of Tertiary sediments. It represents a syncline uniting the Jura Mountains to the High Calcareous Alps (4).
3. *The Prealps*, a very distinctive zone stretching from the Lake of Thun to Lake Geneva and the River Arve. They consist of Paleozoic, Mesozoic and Tertiary formations, entirely foreign to the district in which they stand. They are a travelled mass.
4. *The High Calcareous Alps*, which with their greater height and accompanying glaciers are readily distinguished from (3). The principal peaks which belong to this zone are, from south-

west to north-east, the range of Aravis, the range of the Fis, Mont Buet, Mont Ruan, Tour Saillère and Dents du Midi, the Dents de Morcles, the Diablerets, the Wildhorn, the Wildstrubel, Balmhorn, Doldenhorn, Blümlisalp, Gspaltenhorn, Jungfrau, Mönch, Eiger, Wetterhorn, Titlis, Urirotstock, Pilatus, Tödi, Glärnisch, Churfirsten, Säntis.

South of them rise :

5. *The crystalline Hercynian massifs*, including from southwest to north-east, Mercantour, Pelvoux and Belledonne, Mont-Blanc and Aiguilles Rouges, Gothard, Aar and Gastern-Erstfeld. South of these :

6. *The Pennine nappes*, which form the Ligurian, Cottian, Graian, Pennine and Lepontine Alps.

7. *The Zone of Canavese or the Zone of Roots ;* and finally :

8. *The Calcareous Alps of the South, or Dinarides.*

The Subdivisions of the Eastern Alps. Kober, in his book *Bau und Entstehung der Alpen*, divides the Eastern Alps into the following zones, from north to south :

1. *The Sandstone or Flysch zone*, made up of Cretaceous and Tertiary rocks (Flysch).

2. *The Northern Limestone zone*, well defined, extending from the river Rhine to the river Danube, near Vienna. It consists of sedimentary rocks belonging to the Trias, Jurassic and Cretaceous. The facies of these Mesozoic rocks, characterized by thick masses of limestones, is quite different from the facies of the rocks of the same age found in the High Calcareous Alps of France and Switzerland.

3. *The Grauwacke zone*, the rocks of which are represented by Paleozoic schists and limestones.

4. *The Central zone*, forming the main part of the Eastern Alps. It contains the regions of the Silvretta, Oetztal, Stubaier, Schober, Polinik, and of the Schladminger and Mur Alps. On the crystalline rocks of these mountains we meet with outliers (see p. 16) made up of Mesozoic sediments. On the southern border of these Central Alps occurs the *Drauzug*, with the limestones of the *Karawanken and Gailtaler Alps.*

5. *The Central gneisses*, with their frame of schists, appear as a well-marked zone, discovered long ago, in the Hohe Tauern, from Brenner to Katschberg.

6. *The Karnic main range*, consisting of Paleozoic rocks.

7. *The Southern Limestone Alps* (Dinarides).

Comparisons between the Western and Eastern Alps.
A glance at the geological map, at the end of this book, permits
us to understand the great differences which exist between
these Alpine regions.

Geologically speaking, *the Western Alps disappear underneath
the Eastern Alps.* This is obvious on the right side of the
river Rhine, from the Lake Constance as far as Chur.

On the western side of the river Rhine we notice the ranges
of the High Calcareous Alps ; on the eastern side ranges
made up of limestones appear again, the Northern Limestone
zone of the Eastern Alps, but with facies entirely different
from those of the High Calcareous Alps. The crystalline
massif of the Aar dies out in the Rhine valley, or more correctly
it pitches along with its cover underneath the Eastern Alps.
On the eastern side of the Prätigau (Klosters valley) *the
Tertiary (Flysch) of the Western Alps is overlain by the rocks
of the Silvretta, belonging to the Eastern Alps.*

To sum up, the Eastern Alps, geologically speaking, do not
stand up beside the Western Alps. Nor are the Eastern Alps
the continuation of the Western Alps. *The Eastern Alps are
a gigantic overlap.* This was discovered by Rothpletz. Unfor-
tunately that author affirmed that the movement was, in this
case, directed from east to west. This was a mistake, as we
shall see later on, for the overlap of the Eastern Alps is due
to a movement trending from south to north. *The Western
Alps are overridden by the Eastern Alps*, as was shown splendidly
for the first time by Termier (see Fig. 62.)

To understand the Alps the reader needs some equipment.
In the following chapters of this Introduction I shall devote
some pages to the definition of words used in Alpine tectonic
language. Then the history of the Western Alps will be
sketched, according to the standard work done by Argand.

CHAPTER IV

THE GEOSYNCLINE

Definition. J. Hall, in 1859, pointed out that thick sedimentation and mountain chains were inseparably connected.
Owing to sedimentation a subsidence of the sea floor occurs
and a depression is formed.

J. D. Dana, in 1873, gave the name of *geosyncline* to the
depression in which thick sedimentation occurs, but he considered the formation of the depression as a product of lateral
compression. Thus ridges or *geanticlines*, from which important relief may develop, appear on the bottom of the
geosyncline. Dana called mountains that are made up of
geosynclinal sediments "geosynclinal chains." But this notion
was splendidly worked out later by E. Haug and is at present
one of the basal conceptions of modern tectonics, and, for the
Alps, we may say *the basal conception*.

*A geosyncline is a long marine depression which may be the
result of compression or of stretch* (Argand). *A geosyncline is
situated between two continental masses and is destined to be
filled up by sediments, while geanticlines develop in it.*

The Alpine Geosyncline. Neumayr, before Suess, admitted that a *central Mediterranean* extended, as early as
the Trias, from Asia to Gibraltar across southern Europe,
exceeding the present limits of the Mediterranean. This sea
was *Tethys* or the geosyncline in which the Alps have been
formed, as we shall see later on.

To the north of Tethys we had the southern border of
the great continental mass Eurasia and, to the south,
Gondwanaland. As we are only dealing here with the Alps,
the northern shore—and in this we follow Argand—represents
Europe or the *Foreland* and the southern shore Africa or the
Hinterland.

P. Termier compared the Foreland and Hinterland of the
geosyncline to the jaws of a big vice. In his opinion the
southern jaw approached the northern one. The intervening
geosyncline was thus compressed and folds originated. The
southern jaw even overrode the northern one, and the folds
formed in the geosyncline were thrust against the Foreland.
These folds, according to Termier, were the Alps.

It has been given to Argand, a man of genius, after long
researches in the Alps and detailed mapping, to give us, as
said by E. B. Bailey, "a fascinating interpretation of the
history of the Alps, starting from the first days long before
the great period of the Tertiary movement."

To be able to follow Argand in his interpretation of the
formation of the Alps, let us study:

The Sedimentation in the Alpine Geosyncline. Let us

admit—it will be seen later on that it is a matter of observation—that the Hinterland moved towards the Foreland, leading to the formation of submarine ridges or *geanticlines* in the geosyncline. As shown by Argand these geanticlines are the embryos of the recumbent folds or nappes of the Alps.

Arbenz pointed out that the sedimentation in the Alpine geosyncline depends on the embryonic tectonics. He has established the following types of sediments :

1. Epirogenic sedimentation.
2. Orogenic sedimentation.
3. Thalattogenic sedimentation.

Let us see their characteristics :

1. *Epirogenic sedimentation* takes place on the floor of the epicontinental seas, viz., the seas bordering the continental masses as Foreland and Hinterland (see Fig. 1, Plate I, and page 20). We have to deal here mostly with terrigenous deposits (derived from the continent) or with calcareous deposits formed near the continent in shallow water.

This sedimentation is, moreover, characterized by *cycles of sedimentation,* each cycle showing three different phases :

1. Transgression phase.
2. Inundation phase.
3. Regression phase.

During the *transgression phase* the disintegration of the continental shore takes place with formation of breccias, conglomerates, sandstones. With the *inundation phase* the sedimentation becomes more of a bathyal type. Clays and marls are prevalent. Shallow water deposits characterize the *regression phase*, that may end in emergence.

The cyclical sedimentation is the normal sedimentation in the epicontinental seas. In my opinion these cycles, with phases of transgression and regression of the sea, are due to movements of the continent which are in relation with the orogenic movements in the geosyncline.

2. *The Orogenic sedimentation* is limited to the region of the geosyncline near the shore of a geanticline coming out of the water as a cordillera (see Fig. 1, Plate I and explanation, page 20). In this case clastic (κλαστος broken) material is derived from the geanticline owing to marine erosion. The submarine slope of the geanticline is covered by such debris, and, if it is steep, pebbles and lumps of rock may roll down

to the floor of the geosyncline where they are embedded in deep-sea deposits. If the geanticline does not emerge, but is covered by shallow water—sandstones may be deposited, as well as calcareous sediments formed in shallow water.

3. *The Thalattogenic sedimentation* is characterized by pelagic deposits covering the deepest part of the secondary geosynclines. The cyclical sedimentation does not occur in this type of sedimentation.

Conclusions. Suppose that we have detected several recumbent folds and that our mapping allows us to ascertain their geometrical relations. This is not a goal, but only a first step in modern geology. The second and last step that must be taken will enable us to draw conclusions about the paleogeography. To arrive at this result we have to straighten out the recumbent folds, and our knowledge of the sedimentation in the geosyncline will permit us to detect the parts of the folds representing the epicontinental sea, the secondary geosynclines and the geanticlines.

CHAPTER V

DEFINITIONS

Symmetrical Folds. Strata are generally folded along axes. The simplest fold is a *symmetrical anticline*, in which the strata dip from the axis on either side at the same angle. The converse structure, in which the strata dip in from either side at equal angles to the axis, is known as a *symmetrical syncline* (Fig. 1). The axial plane of a fold contains the axis and makes equal angles with the beds on either side.

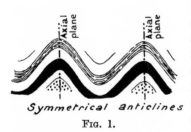

Symmetrical anticlines
FIG. 1.

Unsymmetrical Folds.[1] When the axial plane of a fold is inclined at any angle from the vertical, the fold is said to be *unsymmetrical*. When the axial plane is so much inclined that

[1] Definition after James Geikie (*Structural and Field Geology*).

one limb of the fold becomes doubled under the other, we have the structure known as an *overfold* (Fig. 2). In a fold of this kind the strata which form its lower limb are necessarily turned upside down, and hence the structure is frequently termed an *inversion*. Alpine geologists generally call it the *reversed limb*.

If we suppose that the orogenic push continues, the axial plane approaches horizontality, and we

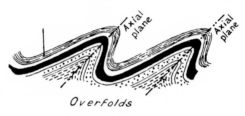

Overfolds

FIG. 2.

have a *recumbent fold,* in which the strata of the inverted limb are stretched. With the continuation of the movement a recumbent anticline may pass into a *thrust mass* or a *nappe*, in which the arch is torn away from the recumbent syncline, riding over the latter (Fig. 3).

Thrusts. Thrusts are dislocations which either :

(1) develop in, and partially replace, the inverted limb of overfolded anticlines, or

(2) allow, without inversion, of the same kind of forward movement as is otherwise characteristic of overfolded anticlines.

The second class of thrusts is often conveniently regarded as an extreme case of the first, and the missing inverted limbs are then described as completely replaced by the thrusts.

All thrusts of the second class are **clean cut.** Wherever a thrust of the first class has produced a relatively important omission in an inverted limb, it also may be described as **clean cut.** In nature, there is no sharp distinction between **clean cut thrusts** and broken inversions. When a thrust mass is based on a clean cut thrust, we have to deal with a *mechanical contact*, as distinguished from an ordinary sedimentary contact.

Nappes. According to Haug, a nappe is *a recumbent anticline* (Fig. 3), the reversed limb of which has partly disappeared owing to stretching. We should thus distinguish between recumbent anticlines and true nappes. As a matter of fact, Alpine geologists nowadays use the term nappe in a general way, that is, also for recumbent anticlines.

In the Alps the great recumbent anticlines developed in the

Alpine geosyncline, and the true nappes are the response of the sedimentary of the Foreland to the push of the recumbent anticlines emerging out of the geosyncline. *A true nappe is thus based on a clean cut thrust of a certain amplitude.*

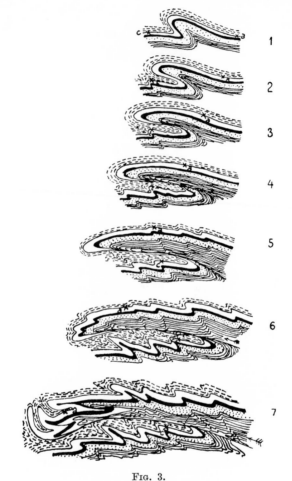

FIG. 3.

1, Overfold. 2–4, Recumbent anticline. 5–7, True nappes. *After Albert Heim (Geologie der Schweiz).*

The nappes of the Pennine Alps, born in the Alpine geosyncline, are recumbent anticlines (Fig. 49). On the other hand, the nappes of the High Calcareous Alps are true nappes (Fig. 4) which rest upon clean cut thrusts.

The frontal part of a true nappe may be divided into several digitations, as in the nappes of the High Calcareous Alps

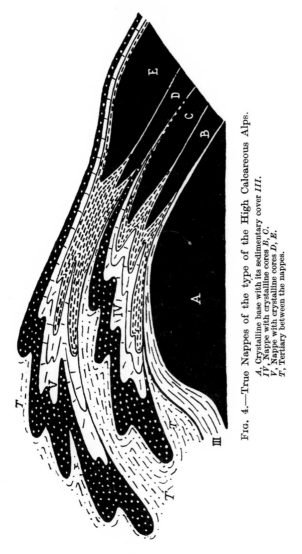

Fig. 4.—True Nappes of the type of the High Calcareous Alps.

A, Crystalline base with its sedimentary cover III.
IV, Nappe with crystalline cores B, C.
V, Nappe with crystalline cores D, E.
T, Tertiary between the nappes.

(Fig. 4), which may produce, on a small scale, recumbent anticlines with reversed limb.

The nappes of the Hinterland are of the type of the nappes of the Foreland, for clean cut thrusts are obvious (Fig. 55).

Roots. The root of a recumbent anticline, or of a nappe, is the core of the anticline in the region where it is more or less vertical and gives the impression of rooting to the depths.

The roots of a nappe or recumbent fold with several digitations are the anticline cores of the different digitations where they are more or less vertical and in connection with the depth. If a nappe, or a recumbent fold, is separated from its roots by erosion we then have, somewhere to the rear, a *zone of roots* (Fig. 7).

Disharmonic Folding. One has to deal with disharmonic folding when in a fold the different strata are not folded in the same way. This occurs, for instance, in an anticline where the outer archbend made of hard limestones gives the general form to the fold, and where the core formed by marls or clays make several folds owing to flowage. Buxtorf has shown that disharmonic folding is well developed in the Jura Range (Fig. 39), and Collet has pointed out in the Morcles nappe, along the Arve valley, that the same phenomenon occurs on a large scale in a great tectonic body. The section (Fig. 29) illustrates how the Jurassic core does not fit into the folds of the Cretaceous cover. We find just the same relation between the folds of the upper Jurassic and those of the Bajocian and Bathonian (Dogger). This disharmonic folding is due, in the one case, to the presence of Infravalanginian marls between the hard limestones of the upper Jurassic and those of the Neocomian and Urgonian, and, in the other, to the presence of the calcareous clays of the Callovian and Oxfordian, resting on hard limestones of the Dogger. The hard strata have a tendency to slide away on the marls or clays and to fold separately, while flowage occurs in the latter. In some cases, in the High Calcareous Alps, the Cretaceous may even be detached from the upper Jurassic and folded separately. A section by Arbenz (Fig. 36) through the High Calcareous Alps of central Switzerland illustrates this phenomenon.

Virgation. Several folds may be grouped so that they form a *bundle of folds*. When such a bundle expands like a sheaf, the different folds turning aside from one another and dying out in the adjacent plain, we have to deal with a *virgation*. This important notion, put forward for the first time by Suess in " The Face of the Earth," has been lately developed and made

more precise by Argand. With this author we recognize two
different kinds of virgations :

1. The unconfined virgation.
2. The confined virgation.

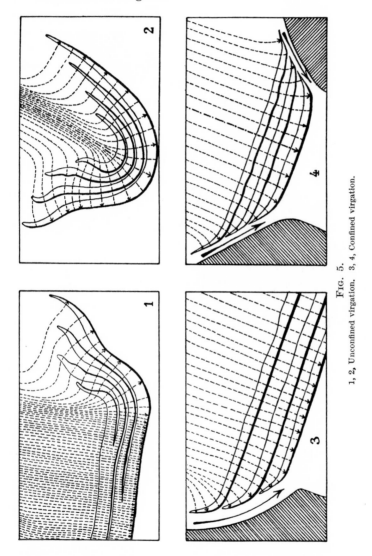

FIG. 5. 1, 2, Unconfined virgation. 3, 4, Confined virgation.

The unconfined virgation (Fig. 5) is recognizable by the
fact that in the central segment of the bundle, with a marked

convexity, the folds press each other close.   At the wing or
wings of the bundle the folds diverge.

The confined virgation (Fig. 5) is due to an obstacle or to
obstacles.   In this case the folds of the central segment of the
bundle have more space between them.   On the other hand,

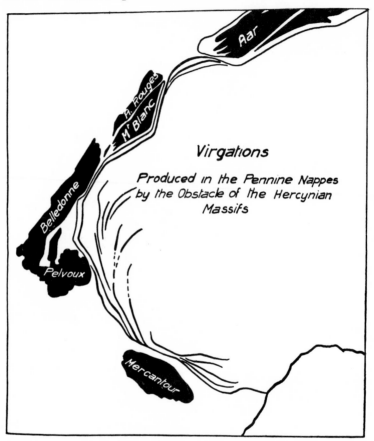

Fᴵɢ. 6. *After E. Argand* (simplified).

the folds of the wing or wings press each other close, so that
their deviated extremities have a tendency to form a straight
line.   We may compare the folds entering a saddle between
two ancient massifs to waves entering a channel in the case of
a *doubly confined virgation,* and to waves rolling obliquely
towards a shore for the *singly confined virgation.*

Window.   A window (Fig. 7) is a cutting due to erosion

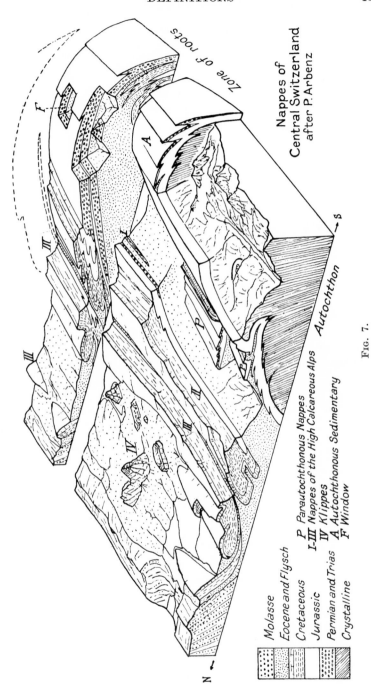

Fig. 7.

Nappes of
Central Switzerland
after P. Arbenz

Zone of roots

Autochthon

Molasse
Eocene and Flysch
Cretaceous
Jurassic
Permian and Trias
Crystalline

P Parautochthonous Nappes
I-III Nappes of the High Calcareous Alps
IV Klippes
A Autochthonous Sedimentary
F Window

into an upper nappe, so that the youngest strata of an underlying nappe appear at the bottom of the valley. The frame
of the window is made up of older strata, belonging to the
normal sequence of the upper nappe.

Windows play a great rôle in Alpine geology. They permit
us to follow underground, along the strike, nappes that
otherwise are so covered by higher elements, that one might
think they had died out. A window is not to be confounded
with an :

Inlier. In the inlier we have to deal with a cutting, due to
erosion, through one or several normal sequences, so that at
the bottom the oldest strata occur.

Nappe-outlier or Klippe. A Nappe-outlier, or Klippe,
is a remnant of a higher nappe spared by erosion. It is
recognizable by the fact that older strata cap younge ones.

Klippes are generally found, in the Alps,
in synclines of a lower nappe, as shown
in Fig. 7. They have been of great help
in reconstructing nappes that have been
almost entirely carried away by erosion,
and also in fixing their limits.

Involution. There are three different
types of involution :

1. The first type is represented by two
nappes of the same age that have been
refolded together later. This type was
discovered by Albert Heim at the
Brigelserhörner (Fig. 8).

FIG. 8.—Involutions.
After Albert Heim
(*Geologie der Schweiz*).

2. Involution takes place when a younger nappe originates
underneath an older nappe and is pressed into the bottom of it
(Fig. 8). The classical example of this case is illustrated by
the involution of the upper nappes of the High Calcareous
Alps underneath the lower ones (Morcles, Diablerets, Wildhorn).

3. The last type occurs when the younger nappe originates
behind the older one and penetrates its back (Fig. 8). This
case is splendidly developed in the " backward folding of the
Great St. Bernard nappe " near Zermatt.

Tear-fault. A tear-fault is due to a horizontal displacement produced across a fold or a bundle of folds. The tear-

fault of the Coin at the Grand Salève, near Geneva, is a very good example of this phenomenon (page 122).

Duplication. A duplication is a repetition, complete or incomplete, of strata belonging to the same stratigraphical sequence. In fact, it is a slice, which overlies the normal stratigraphical sequence, made up of the strata of the latter, or of some few of them. This phenomenon occurs especially in the sedimentary cover of the Aiguilles Rouges-Gastern-Erstfeld massifs, owing to the advance of the Morcles nappe (page 65).

Foundation Folding. We have already seen, in a general way, that the Alps have been formed in a geosyncline, viz., in a marine trough located between two continental masses.

Are all mountain chains of the Earth of the type of the Alps ? The results arrived at in the study of the *Canadian Rockies*, by Collet and Paréjas, show that this is a mountain chain made up of blocks, separated by clean cut thrusts, and pushed over one another. This is quite a different style from that of the geosynclinal folds. But it is also represented in the Foreland of the Alps ; therefore it is necessary to describe it. Argand has named it in French *plissement de fond*, which means a *folding of the foundation*, or a folding which started deep in the foundation of a sedimentary cover. Argand's " plis de fond " has been translated into English by reviewers as " ground-folds." This does not convey Argand's meaning. It is advisable to replace the term " ground-fold " by *foundation folding*. Let us now explain what we mean.

In the *geosynclinal folding* we have to deal with the folding of sediments which fill up a marine trough. When, affected by a tangential push, these sediments fold into *recumbent folds* with an inverted limb, which are characteristic of the geosynclinal folding.

In the *foundation folding* (Fig. 9) the material to be folded is of quite a different nature than in the geosyncline. Indeed, we have to deal with old mountain chains that have been peneplaned and subsequently covered by sediments, after a marine invasion. The rocks forming these old mountain chains are sedimentaries which have been metamorphosed by magmas in the form of batholiths. These rocks—schists, gneisses, granites—are rigid in comparison with the sediments of the geosyncline. When a tangential push makes itself felt on them

s.a.

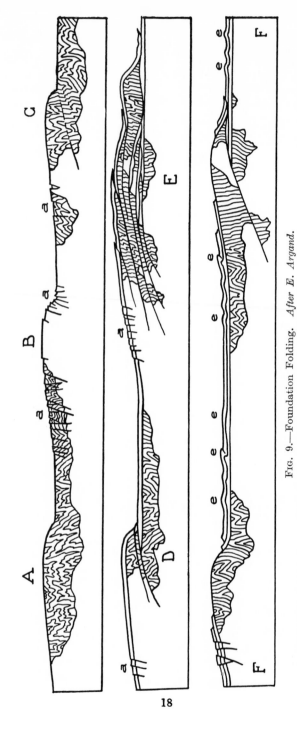

Fig. 9.—Foundation Folding. *After E. Argand.*

A, Anticline due to foundation folding. *B*, Anticline with strike-faults. *C*, Anticline developing into an incipient thrust-structure. *D*, Thrust-mass. *E*, Highly developed thrust-structure, giving rise to the formation of wedges. *F*, Combination of Foundation Folding and folding of the sedimentary cover. *a*, Strike-faults. *e*, folding of the sedimentary cover. The granitic batholiths are represented in white.

18

they cannot fold like sediments, but they break forming huge wedges or blocks thrust over one another and limited by *clean cut thrusts*. The response of the sedimentaries to the breaking up of the foundation is the formation of folds or nappes of the type of the High Calcareous Alps (see Fig. 4).

CHAPTER VI

THE MODE OF FORMATION OF THE ALPS

History. Let us look at the evidence which has led to the conclusion that the Alps are fashioned out of a great pile of recumbent folds or nappes.

Long before Heim, Escher, during the years 1840–1850, conceived the idea of overthrusting. It was not, however, until 1870 that the famous section of the Glarus Double Fold was published with Heim as its warmest adherent. The subject is too familiar to detain us. All that need be said is that in 1884 Marcel Bertrand, following upon his researches in Provence and his reading of Gosselet's description of the Franco-Belgian coalfield, substituted a single for a double fold. Suess in 1892 supported Bertrand's interpretation, and in 1903 Heim at last followed suit.

While discussion still continued regarding the double or single folding of Glarus, Schardt studied the Prealps. During the four years 1890–1893 he realized that in this extensive district Trias is continually to be found overlying Tertiary sediments ; and guided by this relationship, and by a just appreciation of facies considerations, he announced that the Prealps are constituted of a succession of far-travelled thrust-masses of southern origin. Lugeon during 1894–1895 combated this idea, until, won over in 1896 by the accumulated force of his own observations, he began to demonstrate the fundamental rôle of nappes not only in every region of the Alps, but also in the Carpathians, the Apennines, and Sicily. And now to Argand, a pupil of Lugeon, we owe a synthesis of the formation of the Western Alps, based upon a knowledge of the subject such as belongs only to genius.

Argand's Synthesis. Beginning in 1905 Argand, as a result of his researches in Piedmont, was led to study the

Pennine Alps, where he determined the existence of six gneissic recumbent folds or nappes. The investigation was long and difficult, but it threw light on the mode of formation of the whole chain. We shall make use of Argand's sections and explanations (Figs. 1–13, Plate I).

Argand started from the notion of the geosyncline, so splendidly developed by Haug and destined to remain for all time one of the basal conceptions of tectonics. His equipment included a very detailed stratigraphical knowledge, and armed with this he has succeeded in straightening out the recumbent folds, and in thus reconstituting the Alpine region at various stages of its development when the general geosynclinal depression was subdivided by geanticlinal ridges. In the diagram (Fig. 1), showing an Alpine chain in embryonic stage, we may note :

1. The Foreland, with granitic batholiths of Upper Paleozoic (Hercynian) Age.

1'. A swelling at the inner margin of the Foreland determined by the push of the nappes. This swelling we may name the Helvetian Geanticline.

2. The Epicontinental Sea.

3. The Foredeep, or Geosyncline of the Valais.

4. The Frontal Cordillera or Geanticline of Briançon.

5. The Intermontane Depression or Geosyncline of Piedmont.

6. The Second Cordillera or Geanticline of Dolin.

g' shows where Hercynian granites involved in recumbent folds are in part transformed by movement and recrystallization into ortho-gneiss.

b shows the place of origin of certain grits and conglomerates found in the Carboniferous, Permian, Lower Trias, Liasic and Tertiary deposits of the Alps.

b' shows another similar position and illustrates the intrusion phenomena of the greenstones, guided by the dislocations upon which the Geanticline of Dolin advanced.

v' shows how the basic magma of the greenstones escaped at places into the Piedmont Geosyncline, where it consolidated in great measure as pillow lavas.

The diagram (Fig. 1) shows each geanticline of the embryonic Alps as a recumbent fold. Behind the last geanticline should be found what Suess calls the Hinterland of the chain. The

Foreland and Hinterland constitute the boundaries of the great Alpine Geosyncline diversified by its included geanticlines. Together they recall the two jaws of a vice. Termier, director of the " Service géologique de France," has shown how the approach of the two jaws has led to the compression of the geosyncline, and thus to the development of the Alpine chain.   All this will be realized more clearly if we examine Argand's representations of the evolution of mountain systems.

At the close of the *Middle Carboniferous* the beginnings of the two principal nappes show themselves as simple Hercynian anticlines in the great geosyncline (Fig. 2).

In the *Middle Trias* these two embryonic nappes locally rise out of the waters, but the Piedmont depression, 5, continues tolerably deep, as shown by the character of its sediments (Fig. 3).

In the *Lias* (Fig. 4) a strong horizontal push is felt.   The embryos develop, while the depressions deepen.   This is the period of accumulation of the well-known Brèche du Télégraphe (see *b*, Fig. 1).

In the *Middle Jurassic* the Briançon embryo 4 emerges (Fig. 5).

In the *Upper Jurassic* a lessening of horizontal pressure leads to a general slight submergence (Fig. 6).

In the *Middle* and *Upper Cretaceous* the nappes of the Southern Alps, including the upper nappe of the Prealps, start on their journey (their Hercynian cores are marked 8 in Fig. 7, their Mesozoic envelopes 8').

In the *Middle Nummulitic* and *Lower Oligocene* a vigorous push makes itself felt (Fig. 8).

In the *Middle Oligocene* an **orogenic paroxysm** supervenes in which five principal phases may be distinguished :

St. Bernard Phase (Fig. 9).—The Briançon Cordillera makes its main advance, developing into the complex nappe of the Great St. Bernard, 4.   The sediments of the Valais Foredeep are driven forward to yield the nappes of Simplon and Ticino, 3, and also the lower nappes of the Prealps, 3'.   The Foredeep itself is correspondingly displaced and begins to receive the Mollassic sedimentation characteristic of the Swiss Plateau, 3'''.   The Dolin Cordillera passes into the Dent Blanche Nappe, 6.   The upper nappes of the Prealps, 8", begin to separate from their roots and to be carried forward on the backs of the Pennine Nappes, an inclusive term embracing the nappes of Great St. Bernard and Dent Blanche.

Dent Blanche Phase (Fig. 10).—It is now the turn of the Dent Blanche Nappe, 6. The upper nappes of the Prealps, 8″, are concomitantly discharged in front of the Pennine assemblage on the top of the lower Prealpine nappes, 3′ ; from these latter a portion is detached and driven forward to constitute 3″, the external zone, as it is called, of this wonderful group of mountains.

Monte Rosa Phase (Fig. 11).—Born of the Piedmont Geosyncline, the Nappe of Monte Rosa, 5, develops strongly. Concurrently there is considerable back-folding in the Nappe of Great St. Bernard. The Simplon and Ticino nappes, 3, attain their maximum, while beneath them the Helvetian Nappes, 5′, come into being.

Phase of Adriatic Subsidence.—The root region assumes a more or less vertical posture (Fig. 12).

Phase Insubrienne (Fig. 13).—A fan arrangement of the roots is next developed. *The Alps and Dinarides thus face in opposite directions.* The Morcles Nappe and similar parautochthonous folds (8‴ above 1′, Fig. 13) originate beneath the Helvetian Folds (5′, Figs. 11, 12).

The southern jaw of our vice (Indo-Africa of Argand) has approached the northern jaw (Europe) overriding it, and has led to the contraction of the great intervening geosyncline along with the two geanticlines which modified its initial simplicity. These two geanticlines have developed into the two dominant nappes of the Alps, namely, the Nappes of Great St. Bernard and Dent Blanche. To begin with, these two nappes, both of them Pennine Nappes according to current classification, existed alone. As time went on they found themselves moving in a throng. The Great St. Bernard Nappe set in motion the Simplon assemblage, while the Dent Blanche Nappe operated the Monte Rosa group.

The *nappes of the High Calcareous Alps* (based upon 1′, Fig. 13) *owe their existence to the forward drive of the Pennine Alps.*

Kober's Views. Eduard Suess in his *Face of the Earth* advanced the view that *the Alpides*, to which belong the Alps, the Carpathians, the Apennines, the Atlas and the Betic

Cordillera have been pushed towards the north.   The Alpides end in the Balkans and are particular to Europe.

On the other hand *the Dinarides* are a characteristic of Asia and show a southward movement.   To this latter chain belong the Dinarides " sensu stricto," the Hellenides, Taurus and the Himalayas.

According to Suess both mountain chains are morphologically united in the Alps, but tectonically they remain separate units.

We have seen (p. 22) that Argand considers the Alps and the Dinarides to form *a single mountain chain* in which thrusting towards the north is the main feature.   In his opinion an underthrusting at the end of the Alpine building movements produced the thrusting towards the south shown by the Dinarides, which must be regarded as a subsidiary feature.

In the first edition of this book we have dealt especially with the Alps " sensu stricto " and the question of the Dinarides has been laid aside.   Now that we intend to follow Alpine structures in the Western Mediterranean Region it is necessary to draw attention to the views of Kober.

Kober, in 1931, enunciated his principle of the *Orogenide*, viz., a mountain chain with double thrusting, based on Suess's statement that the Alpides are thrust towards the north and the Dinarides towards the south (Fig. 10).

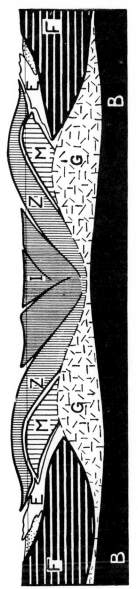

Fig. 10.—Kober's Orogenide or the Theory of the Double Thrusting of the Alpine Chain.

F, Foreland. *E*, Externides (High Calcareous Alps). *M*, Metamorphides (Pennine nappes). *Z*, Centralides (Austrides). *I*, Internides or *Betwixt-mountains. G*, Granite. *B*, Basalt.

Kober's Orogenide includes all the orogenic movements of the Alpine chain.   He divides it into two main groups :
1. A northern thrust-mass.
2. A southern thrust-mass.

Each thrust-mass contains the following elements :
(*a*) The Externides.
(*b*) The Metamorphides.
(*c*) The Centralides.

In the northern thrust-mass the *Externides* are represented by the nappes of the High Calcareous Alps, developed on the Foreland.

The *Metamorphides* are the Pennine nappes formed in the Alpine geosyncline.   They are characterized by the *Schistes lustrés*, which show varying degrees of metamorphism, and the greenstones (ophiolites).

The *Centralides* are the Austrides.

The *Internides* or " Betwixt-mountains " like great islands, characterized by volcanism, divide the Orogenide into two parts.   *Thus to the south of the Internides we find a repetition of the same types of elements as are seen in the northern thrust-mass.*

Kober's latest book, *Das Alpine Europa*, entirely supports this theory, which is not applicable to the Alps " sensu stricto." Indeed, in the Alps the Betwixt-mountains are absent, for the compression was so intense that Alps and Apennines have come together, meeting at a line which is the boundary between the Alps and Dinarides.

Stille and Seidlitz support generally Kober's double thrusting theory of the Alpine Chain.

On the other hand, whilst Staub, in publications, has admitted the possibility of Betwixt-mountains (Zwischengebirge) in certain parts of the Alpine geosyncline proper, he advocates Argand's ideas of a main push towards the European Foreland, with a subsidiary effect towards the south.

The difference between these two theories may not, at first glance, seem great, but is more fully realized on studying the structure of the Western Mediterranean Region.

According to Argand Gondwanaland has been travelling towards Europe, i.e., the Hinterland moved towards the Foreland.   For Kober two Forelands approached one another, viz., the two jaws of the big vice were simultaneously at work ; hence the great thrusting towards the south as well as towards the north.

In the Western Mediterranean Region important thrusting towards the north is known in the Betic Cordillera (South of Spain), in the Balearic Islands, as well as in the Alps. If Kober's views are to be substantiated there should be evidence of an important movement towards the south.

Where is the southern great thrust-mass corresponding to the northern thrust-mass of the Betic Cordillera ? Kober sees it in the mountains of Algeria and Morocco, and the section which he publishes shows it very clearly. Is the evidence in the field so convincing ? Staub does not agree with Kober, and from what I have seen in the field and from what has been published lately by French geologists, especially Moret, I am very sceptical.

When we consider later the Alpi Apuane in Italy (Tuscany), the islands of Corsica and Elba, the question of the direction of movement becomes of great importance. In these regions, as we shall see, whilst nappes are unquestionable, the different authors who have worked at the subject do not agree on the direction of movement. Indeed, there being no visible roots to the nappes, the authors merely adopt the movements which suit either Argand's or Kober's theory.

In the following pages we must consider these two theories as working hypotheses and check them in the field. From what I have seen in different regions of the Western Mediterranean, I regard Argand's theory as fitting the facts much better than Kober's synthesis.

In studying the relations between the Alps and the Apennines we shall have a very good example of the passage from the solid ground of fact to the fragility of speculation.

" Betwixt-mountains " in the Alpine Geosyncline. Argand supposed in his famous paper, " Sur l'arc des Alpes occidentales " (1916), that the Austro-Alpine nappes (Austrides) belong to the Hinterland : that is, to Africa or Gondwanaland. These views had been accepted by R. Staub in his valuable memoir, " Der Bau Alpen " (1924). But during his study of the *Apennines*, he realized that the structure of the Alpine geosyncline is much more complicated than previously admitted, and he pointed out that the Austro-Alpine nappes

(Austrides) *belong to Betwixt-mountains* (in Kober's sense) *and not to the Hinterland.*

Thus the Alpine geosyncline was divided into two parts by *Betwixt-mountains* or great islands, or shoals, made up of Sial. The *Schistes lustrés* were deposited in the northern part of the Alpine geosyncline, whilst in the southern part the sedimentation showed an *abyssal facies.*

Our knowledge of the southern part of the Alpine geosyncline and of the border of the Hinterland being scanty, Staub's results must be considered as preliminary.

Although the Austrides are only " Betwixt-mountains " the idea of the forward drive of the Hinterland towards the Foreland still holds good and cannot be denied.

CHAPTER VII

THE ALPS AND WEGENER'S THEORY

On the basis of Argand's results R. Staub found in the northeastern part of the Swiss Alps the Pennine elements covered by higher nappes belonging more to the type of the " thrust-masses " of the north-western Highlands of Scotland than to the type of the recumbent folds of the Pennine Alps. This series of nappes has been named by Staub the Austrides, for they form the main part of the Austrian Alps.

Staub's " Austrides " represent what Termier and Argand had called, in a much more general way, the Austro-Alpine nappes.

For Swiss geologists the forward drive of the Hinterland towards the Foreland cannot be denied ; it is almost a matter of fact, that is splendidly summarized in Staub's geological map of the Alps.

The intrusion of basic magma in the dislocations upon which the Dent Blanche Nappe advanced (see Fig. 1) is also a matter of fact. I have myself seen the evidence in the field, guided by my friend Argand, and I have shown it to many English geologists.

I must state explicitly that all these results have been obtained independently of Wegener's hypothesis. That is why I think that they are a great support to Wegener's theory.

Conclusion. A southern continent (Africa or Gondwana-land), made of Sial, is separated from a northern continent (Europe or Eurasia), also made of Sial, by a geosyncline or the sea of Tethys of Suess. As a result of the northward drifting of the southern continent, the floor of Tethys thinly coated with Sial has been folded (geanticlines) northwards, thus forming a mountain chain (the Alps). Sima has been injected into the laminated reversed limb of the geanticlines. The southern continent not only encountered the obstacles formed by the northern continent, but its frontal part has been thrust over it.

By the employment of Wegener's ideas we could go a good deal further and—as shown by Argand and Staub—accept a northward drift of Europe, producing a distension to which the Mediterranean is due.

As shown by Lord Rayleigh, many facts make it necessary to abandon Kelvin's classical theory of the cooling of the Earth. The work done by Prof. Joly in this direction is of great value, and it will soon be impossible to resist Wegener's attractive ideas.

CHAPTER VIII

TECTONIC CLASSIFICATION OF THE ALPS

If we now arrange the different parts of the Alps from a tectonic point of view, we arrive at the following groups :

1. The Jura Mountains
2. The Swiss Plateau
3. The High Calcareous Alps
4. The crystalline Hercynian massifs
} The Foreland.

The Pennine Nappes .   The northern part of the geosyncline.
    (Western Alps)
The Austrides   .   .   Betwixt-mountains.
    (Eastern Alps)
The Dinarides   .   .   The southern part of the geosyncline.

In this book I shall follow the tectonic division of the Alps, and the principal parts of it will deal with the Foreland, the northern part of the geosyncline and the Betwixt-mountains.

At our present stage of knowledge it would be preposterous to deal with the southern part of the geosyncline and the Hinterland.

BIBLIOGRAPHY

1. BERTRAND, M.—Rapports de structure des Alpes de Glaris et du Bassin houiller du Nord. Bull. Soc. géol. France, 3ème Ser., T. 12, 1884.
2. SCHARDT, H.—Les régions exotiques du versant nord des Alpes suisses. Bull. Soc. vaud. Sc. nat., vol. XXXIV, No. 128, Lausanne, 1898.
3. HAUG, E.—Les Géosynclinaux et les Aires continentales. Contributions à l'étude des transgressions et des régressions marines. Bull. Soc. géol. France, 3ème Sér., T. 28, 1900, pp. 617–711.
4. LUGEON, M.—Les grandes nappes de recouvrement des Alpes du Chablais et de la Suisse. Bull. Soc. géol. France, 4ème Ser., T. 1, p. 723, 1901–02.
5. LUGEON, M.—Les grandes nappes de recouvrement des Alpes suisses. Congrès géol. internat. Vienne, pp. 477–506, 1903.
6. TERMIER, P.—Les nappes des Alpes orientales et la synthèse des Alpes. Bull. Soc. géol. France, 4ème Sér., T. 3, 1903.
7.—STEINMANN, G.—Geologische Probleme des Alpengebirges. Zeitschr. Deutsch.-Öst. Alpenver, 3f. Bd., 1906.
8. ARGAND, E.—L'exploration géologique des Alpes Pennines centrales. Bull. Soc. Vaud., Sc. Nat., vol. XIV, 1909, pp. 217–276, 3 fig. et 1 pl. de coupes.
9. ARBENZ, P.—Der Gebirgsbau der Zentralschweiz. Verh. der Schweizer. Naturforsch. Ges. 95 Jahresvers., Altdorf, 1912, II Teil, 2.
10. SACCO, F.—Les Alpes occidentales. Turin (Impr. du Collège des Artigianelli) 4º 1913.
11. ARGAND, E.—Sur l'Arc des Alpes Occidentales. Eclogae geol. Helvet., vol. XIV, No. 1, 1916, pp. 145–191, 2 pl.
12. ARBENZ, P.—Probleme der Sedimentation und ihre Beziehungen zur Gebirgsbildung in den Alpen. Festschrift Albert Heim, Zürich, 1919. Vierteljahrsschrift der Naturf. Ges. Zürich. 64 Jahrgang. 1 und 2 Heft.
13. WILCKENS, O.—Allgemeine Gebirgskunde. Jena, 1919.
14. BUBNOFF, S. v.—Die Grundlagen der Deckentheorie in den Alpen. Stuttgart, 1921.
15. HEIM, ALB.—Geologie der Schweiz, 2 vol. 8º, Leipzig, 1919–1922.
16. ARGAND, E.—La tectonique de l'Asie. Congrès géol. internat. Belgique, 1922, p. 171.
17. KOBER, L.—Bau und Enstehung der Alpen. Borntraeger. Berlin, 1923.
18. STAUB, R.—Der Bau der Alpen. Beitr. zur geol. Karte d. Schweiz, N.F. 52e Lief. Bern, 1924.
19. HAUG, E.—Contribution à une synthèse stratigraphique des Alpes occidentales. Bull. Soc. géol. France 4e s. T.25. 1925.

20. COLLET, L. W.—The Alps and Wegener's Theory. Geographical
    Journal, April 1926.
21. SEIDLITZ, W. VON.—Entstehen und Vergehen der Alpen. Stuttgart
    (Enke), 1926.
22. HERITSCH, F.—The Nappe Theory in the Alps. Translated by
    P. G. H. Boswell. London. Methuen. 1928.
23. KOBER, L.—Das Alpine Europa. Berlin, Borntraeger. 1931.
24. CADISCH, J.—Geologie der Schweizer-alpen. Zurich, Beer & Co.
    1934.

GENERAL MAPS

1. Carta geologica delle Alpi occidentali 1/400,000. R. Ufficio geologico. Roma, 1908.

2. HEIM, ALB., und ∫Geologische Karte der Schweiz 1 : 500,000
   SCHMIDT, C. ⌊2e Auflage, 1911. Neudruck, 1927.

3. ARGAND, E.—Les nappes de recouvrement des Alpes occidentales
   Carte structurale. 1/500,000. Matériaux Carte géol. Suisse,
   N. S. Livraison 31, planche 1 (carte spéciale No. 64). Berne, 1912.

4. STAUB, R.—Tektonische Karte der Alpen 1/1,000,000. Beitr. geol.
   Karte d. Schweiz, Spezialkarte No. 105 A. 1923. Profils 3 pl.
   Spezialkarte Nos. 105 B1, 105 B2, et 105 C, 1926.

5. Carte géologique de la France au 1/1,000,000. 3e èd. Paris, 1933.

PART II

THE FORELAND

CHAPTER I

THE DIFFERENT PARTS OF THE FORELAND

As seen above, the main zones of the southern border of the Alpine Foreland are, from south to north :

1. The crystalline Hercynian massifs.
2. The High Calcareous Alps.
3. The Swiss Plateau.
4. The Jura Mountains.

But *the Foreland in itself is made up of the Hercynian Peneplain*.

The High Calcareous Alps, the Swiss Plateau and the Jura Mountains represent its Mesozoic and Tertiary sedimentary cover, more or less folded. On culminations of pitch the sedimentary cover has been taken away by erosion and the crystalline rocks of the Hercynian Peneplain appear.

These culminations of pitch form the following crystalline Hercynian massifs, from south-west to north-east (Plate XII) :

1. Mercantour.
2. Pelvoux and Belledonne.
3. Mont-Blanc and Aiguilles Rouges.
4. Aar, with Gothard to the south and Gastern-Erstfeld to the north.

As pointed out previously, the Aar massif in the Rhine valley pitches underneath the Eastern Alps. Is it possible to think of a prolongation of the crystalline Hercynian massifs underneath the Eastern Alps ?

Three *windows* occur in the Eastern Alps, from south-west to north-east, as follows :

31

1. The window of *the Lower Engadine.*
2. The window of *the Tauern.*
3. The window of *the Semmering.*

Windows generally originate (see page 14) on culminations of pitch. There, the higher nappes having been eroded out, the lower ones are brought to light.

The two first windows show a reappearance of Pennine nappes (Western Alps) underneath the nappes of the Austrides (Eastern Alps). This fact confirms the view expressed above, viz., that the Eastern Alps override the Western Alps.

According to R. Staub, the windows of the Lower Engadine and of the Tauern are due to culminations of the substratum, not only of the Austrides, but even of the Pennine nappes they override. Thus it seems possible that the windows referred to above, with the exception of the Semmering, are a repercussion, at the surface, of culminations of crystalline Hercynian massifs masked by the Pennine and Austride nappes.

If we now consider the window of the Semmering, in which the lower Austride nappes occur, it is not so simple a matter. Indeed, there is in front of this window the block of the Böhmer Wald, a huge Hercynian massif. Thus, the formation of the culmination of the Semmering window may be due to this obstacle which compelled the Austride nappes to go up.

To sum up, *the windows of the Lower Engadine and of the Tauern may be caused by the presence of culminations of Hercynian massifs overridden by Pennine and Austride nappes.*

Let us return to the sedimentary cover of the Foreland. The Hercynian massifs have been encountered by the flow of the Pennine nappes. Being rigid they could not fold, but yielded by breaking up into a series of crystalline wedges, which slid on one another. The sedimentary cover, on the other hand, more pliable, was folded, owing to the formation of the crystalline wedges, into a succession of folds or nappes, to which the High Calcareous Alps are due.

The Jura Mountains consist of a succession of anticlines and synclines. The anticlines are nothing but ridges in the scenery, and the synclines form the valleys. The Jura Mountains do not represent a deep folding, being the outer virgation (see page 12) of the Alps. Indeed, we have to deal here with Mesozoic and Tertiary strata detached and folded on the Middle Trias, owing to the presence of salt beds which play the rôle of a lubricant.

The Swiss Plateau is made up of Tertiary formations, of fresh-water and marine origin, due to the disintegration of the Alps during their formation. The Jura Mountains unite with the High Calcareous Alps, underneath the Swiss Plateau, which plays the rôle of a syncline.

CHAPTER II

THE CRYSTALLINE HERCYNIAN MASSIFS

INTRODUCTION

As it is not possible, in this book, to deal with all the Hercynian massifs referred to above, we shall consider, in general :

1. The massifs of Mont-Blanc and Aiguilles Rouges.
2. The massifs of the Aar, Gastern, Erstfeld and Gothard.

Then the Aiguilles Rouges and Mont-Blanc massifs will be described with more detail, for they are great climbing resorts and can be easily approached by rail along the vale of Chamonix.

As mentioned above, the Mont-Blanc and Aar massifs represent two culminations of the Alpine Foreland. They are thus separated by a saddle or transverse depression. We shall see, later on, that the High Calcareous Alps have been piled up in it. But before dealing with them, let us examine the relations between the Mont-Blanc and Aar massifs.

The massif of the Aiguilles Rouges de Chamonix which, to the north of Mont-Blanc, plays an important orographic rôle, is represented north of the Aar Massif by the Gastern granitic Massif. This latter massif constitutes the lower part of the valley of the same name and the base of the Eiger–Mönch–Jungfrau–Breithorn range.

The Aar granitic Massif thus corresponds to the Mont-Blanc. The trough—zone of Chamonix—dividing the Mont-Blanc from the Aiguilles Rouges, having been subjected to great pressure from the former, is narrower than that of the Lötschental between the Bietschhorn and Gastern massifs. Those familiar with the Lötschental will perhaps object that the rocks there, are entirely different from those which occur at the Mont-Vorassey or at the Col de Balme. This is indeed the case. Paragneisses occur in the Lötschental, and limestones and shales in the zone of Chamonix. This is easily explained, however, as Buxtorf and Collet have shown, by

34

the axial uplift to the north-east. The crystalline schists appearing, in Dr. Paréjas' profiles, as narrow blade-like masses penetrating through the sediments of the zone of Chamonix, constitute the zone of the Lötschental, where these same sediments have been removed by erosion.

CHAPTER III

THE MONT-BLANC AND AIGUILLES ROUGES MASSIFS

GENERAL DESCRIPTION

Tectonics. The Mont-Blanc Massif is separated from the Aiguilles Rouges Massif by the valley of Chamonix, formed by erosion in the sedimentary formations which constitute the *Zone of Chamonix.*

The Massif of Mont-Blanc and that of the Aiguilles Rouges represent, from the geological point of view, remains of the Hercynian chain which formed the southern border of the Foreland of the Alpine geosyncline. The sedimentary rocks of the Hercynian chain have been metamorphosed by a granitic magma with formation of a granitic batholith.

The massifs of Mont-Blanc and Aiguilles Rouges *formed originally one extensive crystalline massif in which the strike of the Hercynian folding was N.–S.*

The approach of the Mont-Blanc Massif to the Aiguilles Rouges Massif was due to the Alpine folding. In fact, this part of the Hercynian chain formed an obstacle to the advance of the Nappes issuing from the Alpine geosyncline. We shall study later, in more detail, the effect of the Alpine folding on these massifs, but we can already state that *the Mont-Blanc Massif was thrust over the zone of Chamonix.*

Paleogeography. During Permian times the Hercynian chain was peneplained, then from Lower Trias to Upper Cretaceous times, it was invaded by the sea. There was a new emersion at the end of Cretaceous and during Early Eocene times, followed by transgression of the Nummulitic sea. On the crystallines of Mont-Blanc and the Aiguilles Rouges we find therefore a sedimentary cover ; but this has been removed by erosion in the higher parts of the Mont-Blanc Massif and we only find it around the lower parts of the Massif.

PLATE II. The Granitic Aiguilles of Chamonix (Mont-Blanc).
The ledge covered by moraines consists of crystalline schists. *Photo by F. Monnier.*

At the summit of the Aiguilles Rouges, however, we still find a sedimentary cover, as we shall see later.

The Aiguilles of Chamonix.   Let us now direct our attention to the chain of the Aiguilles.   From the valley an initial slope brings us to a platform which from Plan de l'Aiguille leads to Montenvers.   The rocks which form this basement are of a sombre tint in contrast to those of the Aiguilles which weather in lighter and warmer hues.   This contrast is due to their being different kinds of rock.   The basement rocks are crystalline schists, whereas granite forms the Aiguilles, the latter being also designated *protogine* (from protogonos).

Central Portion of the Massif.   All the central portion of the Mont-Blanc Massif is composed of granite.   It is indeed this rock which occurs along the ridge from the Aiguille du Midi to the Grands Charmoz.   It is also granite we encounter during the ascent to the Dru, to the Aiguille du Moine, and on the spur which from this peak is directed towards the Aiguille Verte.   The mighty slopes of the Grandes Jorasses, Mont Mallet, and the Aiguille du Géant are of granite.   Further, to the north-east, in the Swiss portion of the chain, the most important peaks are of the same rock, namely, the Aiguilles du Chardonnet and d'Argentière, the Tour Noir, the Darei and the Aiguilles Dorées.

With the aid of a field-glass let us examine Mont-Blanc. We have before us the Grands Mulets and Pitschner's Rock, their dark colour testifying that they belong to the crystalline schists.   This is also true for the Rochers des Bosses.   La Tournette, the last rocky eminence on the flanks of Mont-Blanc, is in the mica-schists, and it is probable that the ice-cap of the summit rests on the same rock.   Thus, in the loftier portions of the chain, the granite is covered by a mantle of crystalline schists, and if the several peaks of the " Aiguilles " are of granite, this is because the crystalline schists have been removed by erosion, being preserved only on the northern slopes, where they are overturned and protected by the granite which rests on top of them.

The Rocks

The rocks of the Aiguilles Rouges and Mont-Blanc massifs have been divided into the following Complexes by Corbin and Oulianoff :

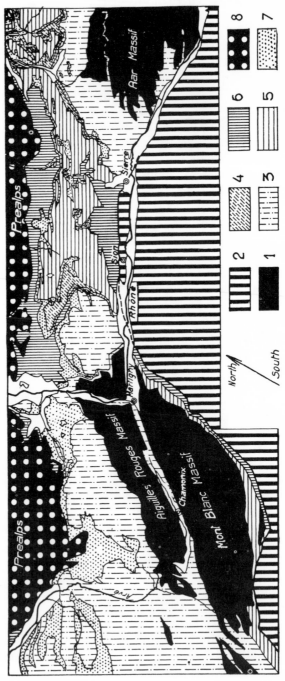

Fig. 11.—Tectonic Map of the Mont-Blanc and Aar Massifs. *After P. Arbenz (with modifications).*
1, Hercynian massifs. 2, The Great, St. Bernard Nappe. 3, Autochthon and Mordes Nappe. 4, The Diablerets Nappe. 5, The Wildhorn Nappe. 6, The Upper Nappes of the High Calcareous Alps. 7, Tertiary Flysch. 8, Prealps.

1. Prarion complex.
2. Servoz—Les Houches complex.
3. Pormenaz complex.
4. Brevent complex.
5. The Aiguille du Goûter and Vallorcine complex.
6. The zone of contact-metamorphism due to granite intrusion.
7. The Granite complex.

Prarion Complex. The principal ridge of Prarion is made up of gneisses almost entirely without quartz of a Pre-Carboniferous age and not of Post-Carboniferous as pointed out by Michel Lévy.

Servoz—Les Houches Complex. This complex is characterized by a monzonitic granite with a cover of mica-schists. These rocks are traversed by acid veins.

Pormenaz Complex. The principal mass of this complex is formed by an alkaline granite with pink felspars. The schists of its cover are very much crushed.

Brevent Complex. The principal rocks are mica-schists with two micas, of which biotite is predominant. When the proportion of felspars increases, the mica-schists pass into gneisses. To the east the schists have been penetrated first by veins of quartz, then by veins of aplite and pegmatite until orthogneisses are met with.

Furthermore, the Brevent complex contains two zones of amphibolites and paleozoic limestones.

This complex crosses the Chamonix valley under the sedimentaries and reappears at Les Rognes, the great scarp separating the Aiguille du Goûter from the Col de Voza. There at the foot of Mont-Blanc the minerals show very distinct deformations due to pressure.

The Aiguille du Goûter—Vallorcine Complex. This complex is characterized by the granite of Vallorcine and by metamorphosed sedimentary rocks. Among the latter hornfels and conglomeratic gneiss are prominent. This complex has been injected by acid rocks.

The Zone of Contact-Metamorphism due to the Granite. This zone is characterized by very numerous apophyses made up of granite, aplite, pegmatite, diorite, syenite, which penetrate, digest, and metamorphose the schists.

The orientation of this zone is given by the glacier des Bossons.

The Granite Complex. Different facies of granite form this complex ; they range from a porphyritic granite with felspars, 10–15 centimetres long, to fine-grained granite passing into microgranite and aplites.

Belts of xenoliths (hornfels–schists) occur. They represent remnants of synclines which have not been entirely digested by the granitic magma. The orientation of these belts is N. 10° E.

Zones of very great mylonitization cut through the granite with a N.E.–S.W. direction. The mylonites are due to the effect of the Alpine folding, as we shall see later.

The Carboniferous. In the Aiguilles Rouges the Carboniferous occurs as conglomerates, sandstones and schists. Conglomerates with large rounded pebbles are well displayed in the region of Lake Pormenaz on the north-eastern slopes of the chain, and on the south side of the valley of the Eau Noire, where they were noticed for the first time by H. B. de Saussure. These Conglomerates of Vallorcine, as they are frequently called, are justly famous for the remarkable conclusions which de Saussure deduced from his study of them. In Swiss territory these conglomerates are well developed in the vicinity of Salvan-Marécottes, to the south of the road to Chamonix. Some of the constituent pebbles attain a diameter of half a metre. In a general way they belong to the rocks of the Aiguilles Rouges Massif.

Near Servoz and the chalets of Moède numerous fossil plants have been found, enabling us to restrict the age of these Carboniferous formations to the Upper Westphalian.

On the north-eastern slope of Pormena, and in the region of Vallorcine and of Salvan, the Carboniferous rests unconformably on the schists of the Aiguilles Rouges.

AGE OF THE MONT-BLANC GRANITE

On the southern side of Mont-Blanc, at the Mont Fréty, the Carboniferous, which is concordant on the crystalline schists, has been metamorphosed. Without doubt this is due to igneous intrusion, and its origin must be sought in the advent of the granitic magma. The Triassic, on the other hand, has suffered no change as can be seen at the Col du

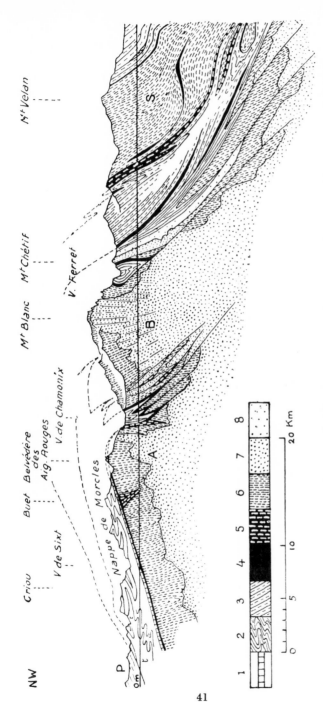

NW

Criou Buet Belvédère V. Ferret M⁺ Blanc M⁺ Chétif M⁺ Velan
        des
V de Sixt Aig. Rouges
        V. de Chamonix

P

Nappe de Morcles

A B S

om t s

1 2 3 4 5 6 7 8

0   5   10   20 Km

FIG. 12.—Section across the Mont-Blanc and Aiguilles Rouges. *After E. Argand.* The zone of Chamonix *after Ed. Paréjas.*
A, Granite batholith of the Aiguilles Rouges. *B,* Granite batholith of the Mont-Blanc. *S,* The Great St. Bernard Nappe. 1, Autochthon (Jurassic-Cretaceous).
2, High Calcareous Alps. 3, Mesozoic of the Pennine Nappes. 4, Trias. 5, Carboniferous. 6, Crystalline Schists. 7, Granite. 8, Quartz-Porphyry.

41

Bonhomme, where its contact with the crystalline schists is a sedimentary contact, or again at the Col de Balme, where it directly supports the crystalline schists.   The intrusion of the granite is thus post-Carboniferous and pre-Triassic, and Lugeon rightly affirms that the Mont-Blanc granite can be but of Permian age.   I will add an observation in confirmation of this ; the crystalline schists of Mont-Blanc never show the effects of weathering, nor the rusty colours of oxidation at their contact with the Triassic.   This proves that the newly born Hercynian mountains were immediately covered by the Triassic sea.   We shall find that this was not the case as regards the Aiguilles Rouges.

THE CHAMONIX ZONE

Description.   The Chamonix zone may be regarded as an erosion groove in the Trias and in the Lias.   It is often deep, and separates the Mont-Blanc Massif from that of the Aiguilles Rouges.   The passes on the north-west side of Mont-Blanc are carved out of the Trias and the shaly Lower Lias of this zone.   They are, from south-west to north-east, the Cols de Truc, Tricot, Mont Lachat and Balme, then that of La Forclaz in Canton Valais.   Summits of little topographical importance flank these notches to the north-west.   They indicate the greater resisting power against erosion possessed by the Middle Lias spathic limestones.   Among these small mountains are the Montagne de Truc, Mont Vorassey, Mont Lachat and the group of the Croix de Fer and Pointe du Van.

Between these summits and the Aiguilles Rouges there is a thick zone of soft Upper Lias shales in which the effects of erosion are evident, as shown in the depression where the village of Bionassey is situated, and also at the Col de Voza. The Chamonix valley is hollowed out of these Upper Lias shales, as in the case of the vale of Martigny-Combe in Valais.

From the presence of the Lias, which was lacking in the sedimentary cover of the Aiguilles Rouges, it is deduced that while the latter was emergent, the Chamonix zone was a marine trough.   Indeed, in the centre of the Chamonix zone the Lias is of a deep-sea facies, which passes towards Mont-Blanc into a shallow-water facies.   Haug has shown that, between Martigny and les Contamines, the Lias of the Chamonix zone belongs to the Dauphiné facies.

Tectonics. The Tectonics of the zone of Chamonix are very important. As shown in Fig. 13, the crystalline rocks which form the substratum appear in the form of two wedges.

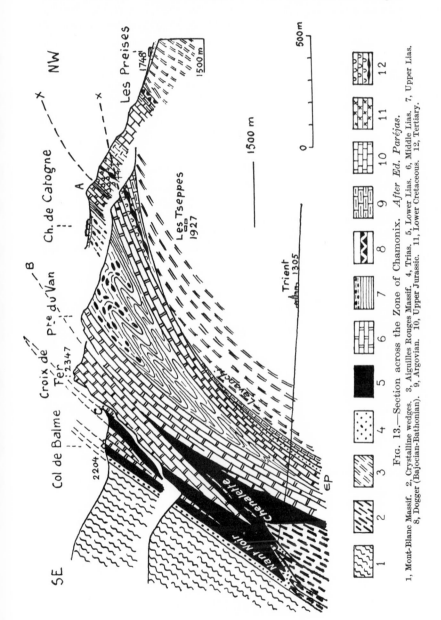

FIG. 13.—Section across the Zone of Chamonix. *After Ed. Paréjas.*

1, Mont-Blanc Massif. 2, Crystalline wedges. 3, Aiguilles Rouges Massif. 4, Trias. 5, Lower Lias. 6, Middle Lias. 7, Upper Lias. 8, Dogger (Bajocian-Bathonian). 9, Argovian. 10, Upper Jurassic. 11, Lower Cretaceous. 12, Tertiary.

We have here still another example of the effect of the Alpine folding on the crystalline rocks which are to be found between the Massif of Mont-Blanc and the Aiguilles Rouges. The crystalline rocks broke, in response to Alpine pressure, whereas the sedimentary rocks forming the cover of these crystalline were folded. The Alpine push was so strong that the sedimentary rocks of the zone of Chamonix were forced out as a nappe which was pushed over the sedimentary cover of the Aiguilles Rouges. This is the Morcles Nappe (Fig. 12).

The Zone of Chamonix forms, therefore, the belt of roots of the Morcles Nappe. At present we regard this belt of roots as being characterized by two anticlines, the cores of which are formed of crystalline wedges.

The Morcles Nappe belongs to the *High Calcareous Alps* which have developed owing to the breaking into wedges of the crystalline of the Hercynian massifs on the southern border of the Alpine geosyncline. We shall study, later, the High Calcareous Alps, but we already arrive at the conclusion that the push exerted on the Hercynian massifs by the nappes issuing from the geosyncline has produced a special type of structure in which one recognizes *foundation folding* (see p. 17) characterized by *clean cut thrusts*. This structure is particular to the Foreland, in opposition to the recumbent folds developed in the Alpine geosyncline.

THE EFFECT OF THE ALPINE FOLDING ON THE MASSIFS OF MONT-BLANC AND AIGUILLES ROUGES

The South-Western Extremity of Mont-Blanc. The response of the Mont-Blanc Massif to these orogenetic forces may now be examined. Being rigid it could not fold, but yielded by breaking up into a series of slices, often wedge-shaped, which slid over one another, while the softer and more pliable sedimentary cover was thrown into a great succession of folds giving birth to the High Calcareous Alps.

In 1897, Ritter recognized in the south-western extremity of Mont-Blanc six of these crystalline wedges, which give a peculiar saw-tooth structure to the Massif. Paréjas has discovered twelve of these wedges over a distance of 5 kilometres between Baptieu and the Col des Fours. The road between Contamines and Col du Bonhomme crosses most of them (Fig. 14).

Some of these crystalline wedges form the substratum of the Chamonix zone and play the rôle of anticline cores.   Four of them are shown in Fig. 13.

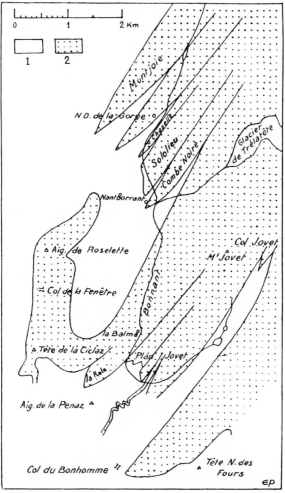

FIG. 14.—Geological Map of the south-west Extremity of Mont-Blanc, showing the Crystalline Wedges.   *After Ed. Paréjas.*

1, Trias and Lias.   2, Crystalline schists.

The Central Portion of the Mont-Blanc Massif.

Corbin and Oulianoff have discovered in the central part of the Massif of Mont-Blanc zones of mylonites in the granite.   These zones of crushed rocks separate granitic wedges which are due, as in the south-west of the Massif, to the Alpine push. The zones of mylonites, being softer and more easily attacked by erosion, play a very great rôle in the present morphology, forming cols or depressions occupied by glaciers. One of these zones passes by the col du Chardonnet, the Col between the Aiguille Verte and the Grande Rocheuse, the Col du Requin, the Col du Midi, the Col du Dôme du Goûter, the Col Infranchissable and continues into the valley occupied by the glacier de Trélatête.

Another of these important zones of mylonites is marked by the glacier du Géant ; it then crosses the northern arete of the Aiguille du Tacul, follows the bed of the glacier de Talèfre, and passes by the Jardin de Talèfre and the col which separates the Droites from the Courtes.

The effect of the Alpine push was, naturally, stronger in the Mont-Blanc Massif than in the Aiguilles Rouges. Schists which lie along the same Hercynian strike show more marked dynamometamorphism to the south of the zone of Chamonix than to the north. The granite of Mont-Blanc is also dynamometamorphosed. Yet, another effect of the Alpine push, demonstrated by Corbin and Oulianoff, is the tectonic contact between granite and schists of the Mont-Blanc massif near Montenvers, above Chamonix. The granite has therefore been thrust over the schists. This contact is clearly seen at l'Angle, near Montenvers, where there are superb examples of flinty crush-rock. In contrast to this, a contact of eruptive nature is to be seen at the base of the Aiguille du Plan and of the Aiguille du Midi.

BIBLIOGRAPHY

MONT-BLANC AND AIGUILLES ROUGES

1. PEARCE, F.—Recherches sur le versant Sud-Est du Massif du Mont-Blanc. Genève, 1898.
2. DUPARC, L., et MRAZEC, L. { Recherches géologiques et pétrographiques sur le massif du Mont-Blanc. Mém. Soc. Phys. et Hist. nat., vol. 33, Genève, 1898.
3. FRANCHI, S.—Escursioni in Valle d'Aosta. Bull. Soc. geol. italiana, vol. XXVI, 1907.
4. LUGEON, M.—Sur l'existence de deux phases de plissements paléozoïques dans les Alpes occidentales. C. R. Ac. Sc. Paris, 30 oct., 1911.

5. LUGEON, M.—Sur quelques conséquences de l'hypothèse d'un dualisme des plissements paléozoïques dans les Alpes occidentales. C. R. Ac. Sc. Paris, 13 nov., 1911.

6. RABOWSKI, F.—Les lames cristallines du Val Ferret et leur analogie avec les lames de la bordure NW des massifs du Mont-Blanc et de l'Aar. Bull. Soc. Vaud. Sc. Nat., 1917.

7. COLLET, L.-W., et PARÉJAS, ED. — Le chapeau de sédimentaire des Aiguilles Rouges de Chamonix et le Trias du Massif Aiguilles Rouges-Gastern. C. R. Soc. Phys. Hist. nat., vol. 37, No. 2, Genève, 1920.

8. OULIANOFF, N.—Quelques résultats de recherches géologiques dans le massif de l'Arpille et ses abords. Eclogae geol. Helvet. XVI, 1920.

9. PARÉJAS, ED.—Géologie de la zone de Chamonix comprise entre le Mont-Blanc et les Aiguilles Rouges. Mém. Soc. Phys. Hist. nat., vol. 39, Genève, 1922.

10. PARÉJAS, ED.—La structure de l'extrémité S.-W. du Mont-Blanc. C. R. Soc. Phys. Hist. nat., vol. 39, No. 1, Genève, 1922.

11. CORBIN, P., et OULIANOFF, N. — Recherches géologiques dans la partie sud-ouest du massif des Aiguilles Rouges (Environs de Chamonix-Servoz). C. R. XIIIe Congrès géol. internat. Belgique, 1922.

12. CORBIN, P., et OULIANOFF, N. — Sur certains caractères du plissement hercynien dans la région Servoz-Les Houches (Vallée de l'Arve). C. R. Ac. Sc., Paris, 28 mai, 1923.

13. OULIANOFF, N.—Le massif de l'Arpille. Matériaux Carte géol. Suisse N.S. Liv. 54, Berne, 1924.

14. PARÉJAS, ED.—La tectonique du Mont-Joly (Haute-Savoie). Eclogae geol. Helv., XIX, No. 2, 1925.

15. CORBIN, P., et OULIANOFF, N. — Deux systèmes de filons dans le massif du Mont-Blanc. C. R. Somm. Soc. géol. France, 1925, p. 202.

16. CORBIN, P., et OULIANOFF, N. — Sur les éléments des deux tectoniques hercynienne et alpine observables dans la protogine du Mont-Blanc. C. R. Ac. Sc., Paris, T. 182, 12 avril, 1926.

17. CORBIN, P., et OULIANOFF, N. — Recherches tectoniques dans la partie centrale du massif du Mont-Blanc. Bull. Soc. Vaud. Sc. nat., vol. 56, No. 217, 1926.

18. CORBIN, P., et OULIANOFF, N. — Les contacts éruptifs et mécaniques, de la protogine et leur signification pour la tectonique du Massif du Mont-Blanc. Bull. Soc. géol. France, 4ème Sér., T. 26, 1926, p. 153.

21. REINHARD, M., und PREIWERK, A. — Ueber Granitmylonite im Aiguilles Rouges Massiv. Verhandl. naturforsch. Gesell. Basel. Bd. XXXVIII, 1927.

22. CORBIN, P., et OULIANOFF, N. — De la différence et de la ressemblance des schistes cristallins des deux versants de la vallée de Chamonix (Massifs du Mont-Blanc et des Aiguilles Rouges). Bull. Soc. géol. France. 4ème sér. T. 27, 1927, p. 267.

23. CORBIN, P., et OULIANOFF, N. — La massif du Prarion et le synclinal complexe de Chamonix. C. R. Ac. Sc., t. 186, 1928, p. 244.

24. CORBIN, P., et OULIANOFF, N. Zones mylonitiques à orientation hercynienne dans le massif du Mont-Blanc. C. R. Ac. Sc., t. 188, 1929, p. 642.

25. CORBIN, P., et OULIANOFF, N. Quelques résultats de recherches géologiques dans le massif de l'Aiguille Verte—C. R. somm. Soc. géol. France, No. 7, 1930, p. 52.

26. CORBIN, P., et OULIANOFF, N. Captures des glaciers sous l'influence de la structure tectonique. C. R. somm. Soc. géol. France, 1931. No. 2, 19 janvier, p. 15.

27. CORBIN, P., et OULIANOFF, N. Carte géologique du massif du Mont-Blanc à l'échelle de 1/20.000e. Feuille *Argentière* avec une notice explicative. Paris, 1932.

28. CORBIN, P., et OULIANOFF, N. Signification tectonique des filons de quartz dans les massifs granitiques. C. R. somm. Soc. géol. France, 1934, No. 8, 23 avril, p. 102.

29. OULIANOFF, N.—Massifs hercyniens du Mont-Blanc et des Aiguilles Rouges. Guide géologique de la Suisse. Wepf. Bâle. 1934.

30. CORBIN, P., et OULIANOFF, N. Carte géologique du Massif du Mont-Blanc (Partie Française). 1:20,000. Avec notices explicatives. (G. Jacquart. Saint-Maur-des-Fossés (Seine).

Feuille Servoz-Les Houches, 1927 ; Chamonix, 1928 ; Les Tines, 1928 ; Vallorcine, 1930 ; Le Tour, 1931 ; Argentière, 1932 ; Mont Dolent, 1934.

BELLEDONNE, PELVOUX, MERCANTOUR

1. LORY, P.—Etudes géologiques dans la chaîne de Belledonne. Note sur la bordure occidentale du massif d'Allevard. Ann. Enseignement sup. Grenoble, T. 5, No. 1. 1893.

2. TERMIER, P.—Le Massif des Grandes Rousses. Paris, 1894. 8º. fig. pl. c. Bull. Serv. Carte géol. Fr. No. 40, T. VI. 1894–95.

3. VIIIe Congrès géologique international. Livret-guide des excursions en France. Paris 1900.

4. SACCO, F.—Sur l'âge du gneiss du massif de l'Argentera. Bull. Soc. géol. France. 4ème s., T. VI., 1907.

5. MICHEL LÉVY, A.—L'Esterel. Etude stratigraphique, pétrographique et tectonique. Bull. Serv. Carte géol. France. No. 130. T. XXI, 1910–11.

6. SACCO, F.—Il gruppo dell'Argentera. Mem. d. R. Acc. d. Sc. di Torino. Ser. II., t. LXI. Torino, 1911.

CHAPTER IV

THE AAR MASSIF

The following are the different parts constituting this massif, from north to south :

1. The granitic massif of Gastern.
2. The Erstfeld gneisses.
3. The paragneisses of the Lötschental.
4. The central zone of Aar granite.
5. The sedimentary syncline of Disentis.
6. The crystalline massif of Tavetsch.
7. The sedimentary zone of the Urseren valley.
8. The crystalline massif of the Gothard.
9. The sedimentary zone of Nufenen.

The Gastern Massif extends from the upper part of the vale of Gastern, above Kandersteg, to the south-west, as far as the upper part of the Engelberg valley, to the north-east, through the upper part of the Lauterbrunnen valley, the base of the Jungfrau, the Urbach valley, the neighbourhood of Innertkirchen, the Gadmen and Wenden valleys.

As in the Aiguilles Rouges of Chamonix, we have to deal here with a granite massif. The granite is exposed in the upper part of the Gastern valley, and in some places on the southern border of the Petersgrat. Elsewhere appear only the crystalline schists of its cover. As in the Aiguilles Rouges, the Trias is resting unconformably on an old land surface of Permian age. That we are dealing with a land surface is inferred from the disintegrated nature of the crystalline schists. The reddish coloration of these schists, underneath the Trias, is due to a pigmentation of iron oxides leached from the land surface.

Carboniferous strata, consisting of slates, conglomerates and sandstones, slightly metamorphosed, have been discovered by Koenigsberger at the *Wendenjoch*, on the southern slope of the

Gastern Massif. This Carboniferous is mechanically overlaid by gneisses which belong to the next internal zone of the Aar Massif, viz., *the Erstfeld gneisses.* It has been followed by Morgenthaler to the south-west, as far as the northern ridge of the Klein Schreckhorn.

The Erstfeld Gneisses are orthogneisses, quite different from the crystalline schists of the cover of the Gastern granite. They are of a greyish hue. Being very hard, they form, to the south of the Gastern Massif, the wildest peaks of the northern border of the Aar Massif, while the Gastern schists occur at the bottom of transverse valleys. The following peaks belong to this zone, from north-east to south-west : Jakobiger, Ruchen, Mäntliser, Krönte, Wichelplank, Murmels- plank, Grassen, Fünffingerstock, Vorder Tierberg, Giglistock, Mährenhorn, Galauistöcke, Hangengletscherhorn, Renfenhorn, Dossenhorn, Ankenbälli, Berglistoçk, Rosenhorn, Mittelhorn, Klein and Gross Schreckhorn.

The granite capping the Mönch, Jungfrau, Rottalhorn, Gletscherhorn, Ebnefluh, forming Kranzberg, Trugberg, Gross Fiescherhorn, belong to the outer part of the Lötschental zone, according to Collet. But as they are the south-westward continuation of Morgenthaler's Erstfeld gneisses, we include them also in this zone.

To sum up, *the Erstfeld gneisses are thrust over the Gastern Massif.*

The Paragneisses of the Lötschental. As seen previ- ously, this zone is equivalent to the Chamonix zone. The best confirmation of this statement is to be found in the range, oriented north and south, which descends from the Balmhorn to the Rhone valley, and forms the Majinghorn, the Faldum and Restirothorn. Indeed, we meet there with remnants of synclines of Triassic and Jurassic rocks on the top of the paragneisses of the Lötschental. Owing to the axis elevation towards the north-east the sedimentary rocks have been taken away by erosion.

Morgenthaler describes, at the southern side of the Erstfeld gneisses, schists which he calls the Fernigen schists. Accord- ing to Collet these schists are nothing but the continua- tion, towards the north-east, of the paragneisses of the Lötschental.

Albert and Arnold Heim have described a very interesting

and important syncline of Mesozoic rocks, occurring at Fernigen in the Meiental (Uri), called the *Fernigen syncline.* This syncline separates the Erstfeld gneisses from the *Central zone of Aar granite.* It has been followed by Morgenthaler farther to the south-west.

A Carboniferous syncline stretches from the Bristenstock to the Faldumalp, through Inschialp, Siglisfadgrätli, south of Fernigen, south-west of Griesenhörnli, Trift Hut, Kilschlistock, Gauli Hut, Gauligrat, south of Ewigschneehorn, Agassizjoch, north of the Lötschenlücke, Goltschenried in the Lötschental.

The Carboniferous synclines (Wenden and Bristen) belong to the Hercynian folding. They are not continuous, having been taken away by the erosion, in many places. The Fernigen syncline, on the other hand, made up of Triassic and Jurassic strata, is due to the Alpine folding.

The Central Zone of Aar Granite has the shape of an ellipse. Its culmination occurs in the region of the Finsteraar-

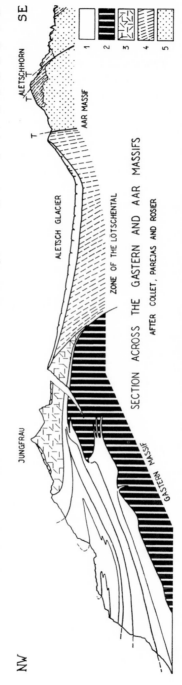

FIG. 15.—Section across the Gastern and Aar Massifs. *After Collet, Paréjas and Rosier.* 1, Sedimentary (Autochthon and Morcles nappe). 2, Crystallines of the Gastern Massif. 3, Granite of the Gastern Massif. 4, Paragneisses of the Lötschental, and Schists of the Aletschhorn. 5, Granite of the Aar Massif. *T,* Thrust-planes in the Aar Massif.

horn. From this peak there is a pitch north-eastward, and the zone dies away at the foot of the Tödi. The region of the Grimsel Pass and of the Rhone glacier belong to it. The great range Aletschhorn-Bietschhorn was carved out of this zone, to the south-west of the culmination. Crystalline schists line the granite on both sides. They may be studied on the left side of the Lötschental and on the right side of the Rhone valley. Good exposures are found in the lower part of the road leading from Naters to Bel Alp and from this locality on the path to the Oberaletsch Hut of the Swiss Alpine Club. Contacts between the granite and the schists are met with in the upper part of the catchment basin of the Kelchbach, above Bel Alp, and also on the way to the Alpine Club Hut when crossing the ridge of the Sparrhorn. At a short distance to the north of the Hut, the contact occurs again.

To get a good idea of the features of the Erstfeld gneisses, the paragneisses of the Lötschental and the central zone of Aar granite, the best plan is to start from the Jungfraujoch and to go down the Aletsch glacier as far as the Concordia Hut of the Swiss Alpine Club.

The gneisses capping the Jungfrau and Mönch are dipping towards the south, and form the Trugberg and Kranzberg, on the southern slope of which they pass under the paragneisses of the Lötschental. The dipping of all gneisses towards the south gives a peculiar feature to the ridges oriented north and south.

At the Concordia Place we are at the junction of three glacial feeders of the great Aletsch glacier. The pass on our right, the Lötschenlücke, is eroded into the paragneisses of the Lötschental. On our left the Grünhornlücke is situated within the border of the granite batholith of the Aar Massif. From the glacier we proceed to the Concordia Hut on the Aar granite, here greatly crushed. In front of us we see the Dreieckhorn on which the contact between the granite and its cover of crystalline schists is obvious.

According to Koenigsberger *the same minerals* are found in the Mont-Blanc granite and in the central zone of Aar granite. This fact shows that the chemical composition of the granitic magmas was very similar.

The Sedimentary Syncline of Disentis. Between Disentis and Truns occurs a syncline made up of Triassic and

PLATE III. The Aar Massif.

On the foreground the summit of the Jungfrau. 1 and 2, Gross and Hinter Fiescherhorn. 3, Agassizjoch. 4, Klein Grünhorn. 5, Grünhorn-lücke. 7, Grüneck. 8, Trugberg.—On the background the great peak is Finsteraarhorn. Peaks 1, 8 and the Jungfrau summit belong to the Erstfeld gneisses (External zone of the Lötschental). Peaks 4–7 and Finsteraarhorn represent the Central zone of Aar granite. *Photo by Ad Astra-Aero.*

Jurassic strata, in which the Lias is absent. These formations represent the sedimentary cover of the southern border of the central zone of the Aar granite. Here, as on the southern side of Mont-Blanc, we notice that the Trias rests conformably on the crystalline schists. This conformity is only apparent. It is not a primary phenomenon, for we have seen that on the top and on the southern border of both Mont-Blanc and Aar Massifs, the Trias rests unconformably on the old land surface of Permian age. The apparent conformity of the Trias with the crystalline schists is then a secondary phenomenon due to the push coming from the development of the Pennine Nappes. This opinion is confirmed by the fact that the Trias is so fully laminated, that in some places the Upper Jurassic comes into contact with the crystalline schists.

The Tavetsch Massif is a crystalline wedge thrust over the syncline of Disentis and, between this locality and the Oberalp Pass, on the central zone of the Aar granite.

The Urseren Zone is very important. Indeed, it stretches over 90 kilometres on the southern side of the Aar Massif, from the Alp Nadèls (Somvix) in the Grisons, to the confluent of the river Massa with the river Rhone, above Naters. The Passes of Oberalp and Furka, as well as the Urseren valley, have been carved out of it.

The sediments of the Urseren zone are different from those of the Disentis syncline. The Permian (continental facies called Verrucano) is very thick ; the clays of the Trias are much more developed than in the Disentis zone. The Jurassic series, with complete Lias, has more likeness to the Pennine *Schistes lustrés* than to the Jurassic of the Disentis zone.

The Gothard Massif, as shown by Niggli, W. Staub and Sonder, is also a part of an Hercynian range. It consists of several anticlinal and synclinal zones, easily recognizable owing to their petrological facies. Indeed, the synclinal zones contain paragneisses, and the anticlinal zones orthogneisses. The Gothard Massif consists of the following zones, from north to south :

1. The Gurschen paragneisses . . . syncline.
2. The Gamsboden gneisses and Cacciola granite . . . anticline.
3. The Guspis paragneisses . . . syncline.

4. The Fibbia gneisses and Rotondo granite . . . anticline.

5. The Sorescia and Tremola series of paragneisses . . . syncline.

These different zones are met with on the road from Hospental to Airolo, over the Gothard Pass.

Schistosity is much more developed in the Gothard Massif rocks than in the central zone of the Aar granite. Indeed, the Gothard Massif has encountered the direct push of the Pennine Nappes. But as the push made itself felt more at the base of this massif, the result is a fan-shaped structure.

The Sedimentary Zone of Nufenen is not a simple syncline as formerly considered. Detailed studies by Eichenberger have shown that it consists of four isoclinal slices (schuppen), made up of Triàssic and Liassic sediments.

The sediments of the Nufenen zone belong to the sedimentary cover of the southern border of the Gothard Massif. The Trias rests unconformably on the crystalline schists. The series of the Lias is complete and may be compared to the formations of the same age of the Torrenthorn (sedimentary cover of the paragneisses of the Lötschental), described by Lugeon. But here the sediments have been metamorphosed and remnants of ammonites (*Arietites*) have been found *in schists containing garnets*.

Eichenberger expresses the opinion that the metamorphism of the Nufenen sediments does not come from magmas of the Gothard Massif, but is due to the pressure of the Pennine Nappes impinging upon this massif. But the Nufenen schists contain tourmaline and it is difficult to explain the presence of this mineral only by dynamometamorphism.

The southern border of the Nufenen zone is overlain by the *Schistes lustrés* of the Pennine Nappes.

Conclusions. The Hercynian Massif of the Aar has been divided into several crystalline wedges owing to the development of the Pennine Nappes at their back. The Erstfeld gneisses, the paragneisses of the Lötschental, the central zone of Aar granite, the Tavetsch Massif, the Gothard Massif, constitute these wedges. It is certain that the latter have also been cut into smaller wedges. We shall see a splendid example of this phenomenon in the Gastern Massif, when studying the Geology of the Jungfrau.

Generally speaking, we may say that the great crystalline

wedges of the Aar Massif, with the exception of the Gastern and Erstfeld Massifs, *represent the roots of the nappes of the*

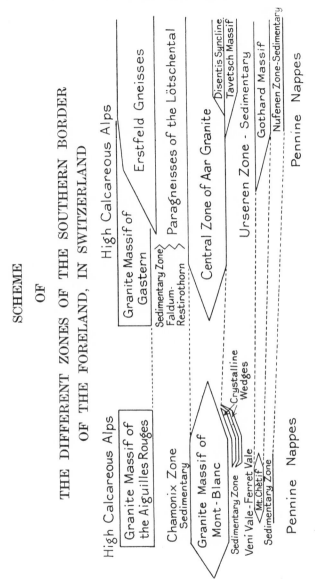

SCHEME

OF

THE DIFFERENT ZONES OF THE SOUTHERN BORDER
OF THE FORELAND, IN SWITZERLAND

High Calcareous Alps, as is splendidly demonstrated by Buxtorf, Lugeon and Weber.

BIBLIOGRAPHY

1. WEBER, F.—Über den Kali-Syenit des Piz Giuf und Umgebung (östl. Aarmassiv). Beitr. z. geol. Karte d. Schweiz. N.F. 14e Lief (44). Bern, 1904.
2. FISCHER, O.—Über einige Intrusivgesteine der Schieferzone am Nordrand des Zentralen Granites, aus der Umgebung der Sustenhörner (Mittleres Aarmassiv). In. Diss. Zürich-Wien, 1905, 8°, 68 p.
3. KOENIGSBERGER, J.—Die kristallinen Schiefer der zentralschweizerischen Massive und Versuch einer Einteilung der kristallinen Schiefer. Congr. Géol. Suède, 1910, pp. 639–671.
4. TRUNINGER, E.—Geologisch-petrographische Studien am Gasterenmassiv. (Inaug. Dissert.), Bern, 1911, 8°, 97p. pl.
5. NIGGLI, P.—Die Chloritoidschiefer und die sedimentäre Zone am Nordostrande des Gotthardmassives. Beitr. z. geolog. Karte d. Schweiz. N.F. 36e Lief. (66). Bern, 1912.
6. STAUB, W.—Beobachtungen am Ostende des Erstfeldermassivs. Geol. Rundschau, vol. III, 1912.
7. NIGGLI, P., und STAUB, W. Neue Beobachtungen an dem Grenzgebiet zwischen Gotthard und Aarmassiv. Beitr. z. geol. Karte d. Schweiz N.F. 45e Lief. (75). Bern, 1914.
8. KOENIGSBERGER, J.—Zur Abtrennung des Erstfelder-vom Aarmassiv und ergänzende Beobachtungen im Aarmassiv. Eclogae Geol. Helvet., vol. XIII, No. 2, 1914.
9. HEIM, ALB., und HEIM, ARN. Die Juramulde im Aarmassiv bei Fernigen (Uri). Vierteljahrsschr. d. Naturforsch. Gesellsch., Zürich. Jahrg. 61 (1916), pp. 503–520.
10. SWIDERSKI, B.—La partie occidentale du Massif de l'Aar entre la Lonza et la Massa. Mat. Carte géol. Suisse, Liv. 47. Berne, 1919.
11. MORGENTHALER, H.—Petrographisch-tektonische Untersuchungen am Nordrand des Aarmassivs. Eclogae geol. Helvet., vol. XVI, No. 2, 1921.
12. HUTTENLOCHER, H.—Vorläufige Mitteilung zur Petrographie und Geologie des westlichen Aarmassivs. Mitteilung d. Naturf. Ges. Bern, 1921, H. 1.
13. HUGI, E.—Das Aarmassiv, ein Beispiel alpiner Granitintrusion. Verh. Schweiz. Naturf. Gesellsch. Bern, 1922. II Teil, p. 86.
14. HEIM, ALB. (nach WEBER, F.).—Ostende des Aarmassivs. Geologie der Schweiz. 2 II. Nachträge, pp. 932–938. Tauchnitz, Leipzig, 1922.
15. EICHENBERGER, R.—Geologische und petrographische Untersuchungen am Südwestrand des Gotthardmassivs (Nufenengebiet). Eclogae Geol. Helvet, vol. VIII, No. 3, avril 1924, p. 451.
16. SCABELL, W.—Beiträge zur Geologie der Wetterhorn-Schreckhorn-Gruppe. Beitr. z. geol. Karte d. Schweiz. N.F. 57e Lief. III (87). Bern, 1926.
17. ROHR, K.—Stratigraphische und tektonische Untersuchung der Zwischenbildungen am Nordrande des Aarmassivs (zwischen Wendenjoch und Wetterhorn). Beitr. z. geol. Karte d. Schweiz. N.F. 57e Liefer. I (87). Bern, 1926.

18. PFLUGSHAUPT, PAUL.—Beiträge zur Petrographie des östlichen Aarmassivs (Petr.-Geol. Untersuch. i. geb. d. Bristenstockes). Schw. Mineral. u. Petr. Mitt. VII, 2, 1927, pp. 322–378.
19. COLLET, L. W., [La Géologie du Hochenhorn. Eclogae geol. et PARÉJAS. [Helvet., vol. XXII, No. 1, p. 61, juin, 1929.
20. WINTERHALTER, R. U.—Zur Petrographie und Geologie des östlichen Gotthardmassivs. Schw. Mineral. Petr. Mitt., X/1, 1930, pp. 38–116.
21. ROSIER, GEORGES.—Contribution à la géologie de l'Aletschhorn (avec une carte). Eclogae geol. Helvet., vol. XXIV, No. 1, pp. 83–124, 1931.
22. WYSS, RUD.—Petrographisch-geologisch Untersuchungen westlich der Grimsel im Finsteraarhorn-Lauteraarhorn gebiet. Mitt. Naturf. Ges. Bern, 1932, in 8°.
23. HUGI, EMIL.—Das Aarmassiv. Guide géologique de la Suisse, 1934, p. 130. Wepf, et Cie, Bâle.
24. NIGGLI, PAUL.—Das Gotthardmassiv. Guide géologique de la Suisse, p. 139, 1934. Wepf et Cie, Bâle.

CHAPTER V

THE HIGH CALCAREOUS ALPS

INTRODUCTION

The High Calcareous Alps consist of sediments which were deposited in an epicontinental sea, on the peneplained Hercynian range forming the southern border of the Foreland of the Alpine geosyncline. The epicontinental sea is, in other words, the shallow-water region of the northern margin of the Alpine geosyncline, as shown by Argand in Fig. 1, Plate I.

Owing to the development of the Pennine Nappes in the Alpine geosyncline, the bottom of the epicontinental sea was uplifted or affected by subsidences. These movements of the Foreland are marked in the stratigraphical sequence of the High Calcareous Alps by facies representing cycles of sedimentation. It is what has been called by Arbenz an *epirogenic sedimentation*. The different phases of each cycle are : (1) Transgression phase. (2) Inundation phase. (3) Regression phase (see page 7). Thus we have here another example of the importance of detailed stratigraphical studies to ascertain the paleogeography of a region.

The High Calcareous Alps are of paramount interest, because the first detailed studies on the mechanics of nappes were made in this region of the Alps by Lugeon.

At present we know that the nappes, out of which the High Calcareous Alps have been carved, are a repercussion of the formation of the Pennine Nappes. These nappes, that we shall study in detail later on, are not nappes of the style of the Pennine Nappes. *They are not recumbent folds, they are overthrusts,* in the sense used by Peach and Horne in their great work on the North-West Highlands.

The nappes of the High Calcareous Alps are overthrusts because they have resulted through the formation of the great crystalline wedges derived from Hercynian Massifs of the type of Mont-Blanc and Aar-Gothard.

When using the word *nappe*, we must always remember the difference in the genesis of the two kinds of nappes, the nappes of the High Calcareous Alps due to the formation of crystalline wedges in the Hercynian Foreland, and the Pennine Nappes born in the Alpine geosyncline. The latter are *recumbent folds*.

It was in the High Calcareous Alps that in 1901 Lugeon arrived at the important notion of the " déferlement vers le Nord," viz. that *the farther a nappe spreads out to the north, the farther it roots to the south*. This rule has, at present, very many exceptions, but played a great rôle during the geological exploration of the Alps.

THE AUTOCHTHON

The Autochthon includes the sedimentary cover of the Aiguilles Rouges, Gastern and Erstfeld, as well as the massifs themselves. The name was given when, on the discovery of the nappes of the High Calcareous Alps, these sediments of the Foreland were supposed not to be folded. But recent exploration of the Aiguilles Rouges, Gastern and Erstfeld Massifs has furnished evidence that the so-called Autochthonous formations are also affected by movements, of a special kind. They show folds and duplications, with repetitions of slices of normal series, and even *clean-cut thrusts*. Several Swiss geologists limit the word *Autochthon* to the *undisturbed* sedimentary cover of the Aiguilles Rouges and Gastern Massifs, whilst they use the word *Parautochthon* for the folds and duplications.

The Mesozoic Cover of the Aiguilles Rouges has been removed by erosion in the higher parts of the massif. There remains of it a single patch, on the highest summit of the chain : the Belvédère (2,966 m.), as was shown by Dolomieu, Necker and Alphonse Favre.

Alphonse Favre considered the sedimentary remnant of the Belvédère to be a normal series from Trias to Jurassic. As he made the ascent of this mountain by its northern ridge, he was stopped by the deep cleft near the summit, and was in consequence unable to examine its southern side, where Collet and Paréjas found, in 1920, various tectonic complications, the meaning of which will be understood later.

Trias. A tract of Triassic sediments overlies the Carbon-iferous on the north-west side of the Aiguilles Rouges. It may be followed from the chalets d'Ayère to the base of the Col d'Anterne, in the bed of the Souay torrent. From the Col d'Anterne Hotel, the Trias forms a ledge leading to a point above the chalets of Villy, and thence to the Col de Salenton. Farther on, and still following the same formation, which is easily recognizable at a distance, owing to the contrast of its yellow colour with the black of the overlying beds, one may reach the vale of Salanfe, at the foot of the Dents du Midi, and

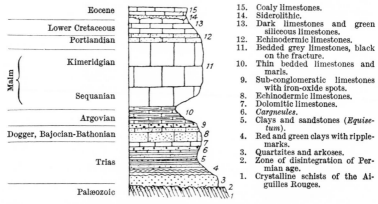

Eocene	15. Coaly limestones.
	14. Siderolithic.
Lower Cretaceous	13. Dark limestones and green siliceous limestones.
Portlandian	12. Echinodermic limestones.
	11. Bedded grey limestones, black on the fracture.
Kimeridgian	10. Thin bedded limestones and marls.
	9. Sub-conglomeratic limestones with iron-oxide spots.
Sequanian	8. Echinodermic limestones.
	7. Dolomitic limestones.
Argovian	6. *Cargneules.*
	5. Clays and sandstones (*Equise-tum*).
Dogger, Bajocian-Bathonian	4. Red and green clays with ripple-marks.
	3. Quartzites and arkoses.
Trias	2. Zone of disintegration of Per-mian age.
	1. Crystalline schists of the Ai-guilles Rouges.
Palæozoic	

(Malm: Kimeridgian, Sequanian)

FIG. 16.—Stratigraphical Sequence of the Autochthon (Sedimentary cover of the Aiguilles Rouges). *After Ed. Paréjas.*

the Col of Jorat, which leads into the Rhone valley. The route passes through a series of depressions of which the principal are : the Col of Vieux, the Col of Barberine and the Col of Emaney. These features show that the Trias of the Aiguilles Rouges is a soft formation.

From the chalets of Villy to the Col of Jorat, the Trias rests unconformably on vertical crystalline schists. In fact, the Trias covers an old land surface of the Permian age. That we are dealing with a land surface is inferred from the disintegrated nature of the crystalline schists. The reddish coloration of these schists, near the Trias, is due to a pigmentation of iron oxides levigated from the land surface.

Jurassic and Cretaceous. During the Lias, the Aiguilles

Rouges were emergent, as is inferred from the total absence of Liassic deposits. The echinodermic limestones, with Ammonites, of the Croix de Fer show that the Bajocian sea invaded the continent. A further stratigraphical gap during Callovian-Oxfordian times points to another emergence followed by an Argovian transgression. The base of the Argovian is marked by coarse lumpy conglomeratic limestones, with yellow spots due to inclusions of ankerite.[1]

The Upper Jurassic limestones (Malm) lie on the Upper Argovian marls. They point to a deepening of the sea, which continued into the Lower Cretaceous.

Tertiary. The Middle and Upper Cretaceous are absent, and the Eocene Siderolithic lies on the Berriasian. Paréjas informs us that oolitic iron-ores were formerly worked on the slopes of the Tseppes and of the Pointe of Carraye, on the western side of the Trient vale. The iron-works were situated below the village of Trient, at a spot called La Ferrère. Blackish carbonaceous shaly marls with bedded limestones occur in the ravine of Catogne : most likely they represent the Priabonian (Fig. 16).

THE AUTOCHTHONOUS SEDIMENTARY AND THE FORMATIONS OF THE JURA MOUNTAINS

The Autochthonous sedimentary unites with the sediments of the Jura Mountains, underneath the Swiss Plateau. But the facies of the formations of the Jura Mountains are quite different from those of the Autochthon, though, in some cases, the intervening distance may be small. The best evidence is to be seen near Geneva, on French territory, where the Salève (see page 119) shows the Jura Mountains facies, and the ranges of the Genevois (Haute Savoie), at only a few kilometres distance, consist of rocks of the High Calcareous Alps.

The oldest post-Hercynian sediments of the Jura Mountains belong to the continental Permian, as is the case for the Autochthon. Then a marine transgression took place, in both regions, at the beginning of the Trias. This stage of the Jura Mountains consists of the three elements of the German Trias (see page 126). In the Autochthon three divisions are also met with, but they correspond to a sedimentation which took place much nearer to the coast.

[1] Iron-bearing dolomite $(CaCO_3(Mg,Fe)CO_3)$.

During the Lias, the Aiguilles Rouges and Gastern were a land surface, while, in the Jura region, black shales with Ammonites were deposited.   As we go up the Jurassic series of the Jura Mountains, we get evidences that the sea becomes more and more shallow till we arrive at the end of the Jurassic period when an emergence is evidenced by the presence of the fresh-water beds of the Purbeck.   On the other hand, the Autochthon shows a Bajocian transgression separated by a Callovian-Oxfordian emergence from an Upper Jurassic and Lower Cretaceous period of comparatively deep sea.   During the Eocene an emergence made itself felt in both regions, and the sediments of the Eocene are continental.   They are called " Siderolithic," owing to their containing iron-stone produced by the disintegration of Upper and even Middle Cretaceous strata.

To sum up, the bathymetrical features of the sea of the Jura Mountains were quite different from those of the Autochthon.   Except during the Eocene, an emergence of one of the regions was balanced by a submergence of the other.   We thus arrive at the result that the Autochthonous sedimentary was deposited on a ridge, a kind of geanticline, of the Hercynian Foreland, separating the basin of the Jura Mountains from the Alpine geosyncline.   This geanticline, called by Haug the *Helvetian Geanticline*, is shown in Fig. 1, Plate I, and page 20.

THE NAPPES

Introduction.   The response of the Hercynian Massifs to the orogenic forces set up in the Alpine geosyncline, was the formation of crystalline wedges which slid over one another, while the softer and more pliable sedimentary cover was thrown into a great succession of nappes, giving birth to the greatest part of the High Calcareous Alps.

Divisions.   Lugeon has shown that the High Calcareous Alps are made up of nappes which override the Autochthon (Fig. 17 and 18).   From top to bottom they read as follows :

VI.   The Oberlaubhorn Nappe.
V.   The Mont Bonvin Nappe.
IV.   The Plaine Morte Nappe.
III.   The Wildhorn Nappe.
II.   The Diablerets Nappe.
I.   The Morcles Nappe.

Owing to the wedging in of these nappes between the Massifs of Mont-Blanc and Aar they have been piled over one another in Valais on the right bank of the Rhone. The roots of the Morcles Nappe were discovered by Paréjas in the Chamonix

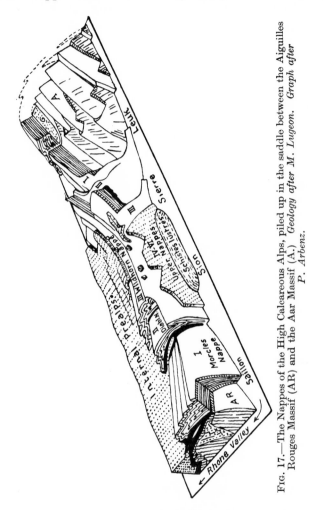

Fig. 17.—The Nappes of the High Calcareous Alps, piled up in the saddle between the Aiguilles Rouges Massif (AR) and the Aar Massif (A.) Geology after M. Lugeon. Graph after P. Arbenz.

zone, and the roots of the Diablerets Nappe should be found in the Mont-Blanc Massif, and those of the Wildhorn Nappe on its southern slopes. Rabowsky drew attention in 1917 to the presence on the southern slopes of Val Ferret of one and sometimes two wedges of crystalline rocks composed of porphyries very much crushed, of aplite and of crystalline schists. He has followed them for 16 kilometres to the Petit Col Ferret, and they are certainly continued on the Italian side. They override the sedimentary covering of Mont-Blanc and very probably represent the roots of the Wildhorn Nappe. The granite and porphyry wedges of Mont Chétif probably belong, in the opinion of Rabowsky, to the upper nappes (IV–VI).

As the High Calcareous Alps run across Switzerland, we shall have to correlate the classical nappes of the northern side of the Rhone valley with the nappes recorded by Oberholzer, Albert Heim, Arnold Heim, Arbenz and Buxtorf between the Aar river and the Rhine river. This will not be easy owing to the fact that the Aar Massif, though it is the tectonic equivalent of the Mont-Blanc Massif, is more complicated.

From a stratigraphical point of view, the upper nappes (III–VI) are characterized, especially in the Cretaceous and Tertiary, by formations deposited in deeper water than was the case for the lower nappes. Indeed, this fact is easily comprehensible if one takes into account that the higher nappes root farther to the south than the lower ones. The sediments of the highest nappe must then represent a passage to the *Schistes lustrés* of the northern part of the Alpine geosyncline.

Let us now study :

The Relative Age of the Nappes. When dealing with this subject, phenomena of involution (see page 16) play a rôle of paramount importance. A glance at Fig. 18 enables us, better than a long description, to understand why nappes IV–VI are the oldest. Their involution is clear, and was due to the subsequent development of nappe II (Diablerets) and of nappe III (Wildhorn) underneath them. Nappe I (Morcles) is youngest of all. It was the last to advance, as it is enveloped by the upper nappes (IV–VI). The intercalation of the Morcles Nappe between the Autochthon and the Diablerets Nappe, will help us to explain several tectonic features of the Autochthonous sedimentary.

TECTONICS OF THE AUTOCHTHON

The tectonics of the Autochthon show the following features :

Clean-cut thrusts, due to small crystalline wedges. This phenomenon is a repercussion of the formation of the Pennine

Nappes. The best examples of it are met with at the base of the Jungfrau, when going up from Stechelberg (above Lauterbrunnen) to the Rottal Hut of the Swiss Alpine Club, as will be explained later on (see p. 70).

Duplications of the Mesozoic rocks, caused by the travelling northwards of the Morcles Nappe. In this case, as shown by Lugeon, the Autochthonous sedimentary has been peeled off the crystalline rocks on small culminations and piled up against obstacles or accumulated in depressions of the surface of the Aiguilles Rouges and Gastern Massifs.

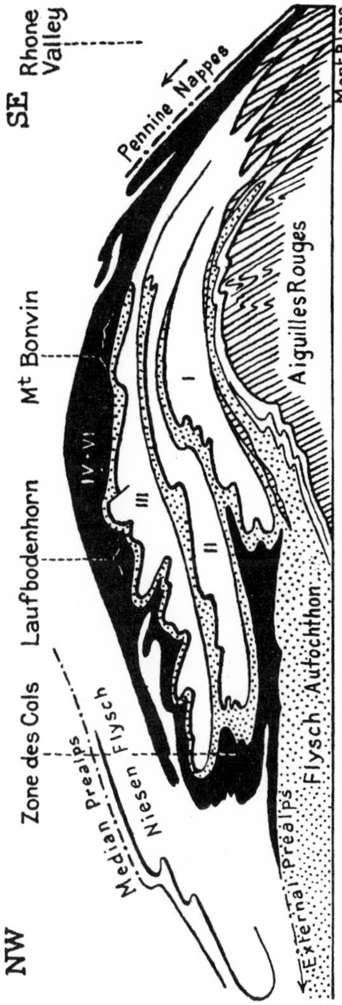

The sedimentary cap of the Belvédère, the highest summit of the Aiguilles Rouges of Chamonix, supplies a splendid example of this structure as shown by Collet and Paréjas. Let us study this section (Fig. 19).

The normal sequence of the Trias in the Aiguilles Rouges is from bottom to top: (1) Quartzites and arkoses. (2) Red and green clays. (3) Dolomitic limestones with intercalations of black shales with *Equisetum*.

When climbing up the Belvédère, on the southern ridge,

S.A.

Fig. 18.—Generalized section across the High Calcareous Alps, based on Lugeon's Scheme. After Albert Heim (Geologie der Schweiz).

1, Morcles Nappe. II, Diablerets Nappe. III, Wildhorn Nappe. IV–VI, Upper Nappes (Plaine Morte, Mont Bonvin, Oberlaubhorn).

one meets with the quartzites and arkoses, marking the unconformity of the Trias on the crystalline schists. But the red and green clays are absent, and the upper formations of the Triassic series rest, strongly folded, on the flat quartzites and arkoses. If one walks along the ledge formed by the quartzites, it will be noticed that, half-way to the northern ridge, the Trias is complete. This shows that on the southern ridge the strata of the Upper Trias have slid over the red and green clays, which played the rôle of a lubricant, and have been locally folded. Owing to this forward movement the red and green clays have been laminated to such an extent that they are absent in places. Moreover, the folds of the Upper Trias are overridden, at the summit, by a slice of Upper Jurassic. This is a duplication. Farther to the north, as shown in the section, echinodermic limestones of the Bajocian occur underneath the slice of Upper Jurassic. One might think, on studying only the sequence on the northern ridge, that a normal series there connects the Trias with the Upper Jurassic. But if one starts on the southern ridge, where an obvious thrust separates the Upper Trias

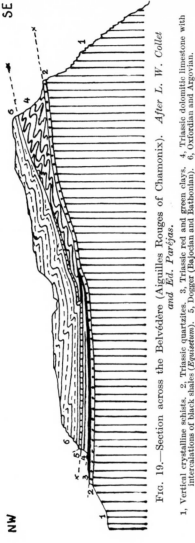

FIG. 19.—Section across the Belvédère (Aiguilles Rouges of Chamonix). *After L. W. Collet and Ed. Paréjas.*

1, Vertical crystalline schists. 2, Triassic quartzites. 3, Triassic red and green clays. 4, Triassic dolomitic limestone with intercalations of black shales (*Equisetum*). 5, Dogger (Bajocian and Bathonian). 6, Oxfordian and Argovian.

and the Upper Jurassic, one realizes that this thrust-plane can be followed as far as the northern ridge. The normal sequence there is only apparent, for a thrust-plane separates the Upper Trias from the Bajocian.

These structures are caused by the friction exerted on the Autochthon by the northward travel of the Morcles Nappe.

The Accumulation of Folds of the Autochthonous Sedimentary in front of the Morcles Nappe. In this case the Autochthonous sedimentary has been detached from its base by the frontal part of the Morcles Nappe, and pushed north-

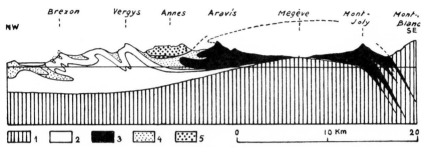

Fig. 20.—The Accumulation of Folds of the Autochthonous sedimentary in front of the Morcles Nappe. *After Ed. Paréjas.*

1, Hercynian Massifs.   2, Autochthonous sedimentary.   3, Morcles Nappe.   4, Tertiary.
5, Nappe-outlier (Klippe) of the Annes.

wards, where it forms a bundle of folds (Fig. 20). This phenomenon is a *décollement* exemplified, as shown by Paréjas, by the ranges of the Genevois (Haute Savoie), on the way from Geneva to Chamonix.

The Geology of the Jungfrau [1]

From the Lauterbrunnen valley up to the summit of the Jungfrau we find the greatest vertical cliff, of about 3,200 metres, in the Swiss Alps. If this mighty cliff were the only exposure of the geology of the Jungfrau, we would know little about its structure. Fortunately a valley cuts through the mountain on its western side, it is the Rottal (red valley). A path leads from the village of Stechelberg, 6 kilometres above Lauterbrunnen, to the Swiss Alpine Club hut (2,700 m. high),

[1] Geological map : Chaine de la Jungfrau 1 : 25,000, by L. W. Collet and Ed. Paréjas, Carte géologique spéciale 113. Francke. Berne, 1927.

which is a starting-point for the ascent of the Jungfrau. It is along this route that the Geology must be studied, following the profile published by Collet and Paréjas (Fig. 21). If one wants to avoid climbing the Jungfrau, the whole profile may be studied with the binoculars from Ober Steinberg, a small hotel about two hours' distance from Stechelberg. Let us now examine the Jungfrau section (Fig. 21), which will allow us to study the relations between the Morcles Nappe and the Gastern Massif.

This section shows three well-defined parts, in ascending order :

A. *The foundation*, made up of crystallines of the Gastern Massif.

B. *The limestones*, very thick on the north-western side and thinning out towards the south-east.

C. *The granite of the Jungfrau* resting on the limestones and forming the summits of the Jungfrau and Rottalhorn.

The Foundation (Gastern Massif)

From a structural point of view, the crystallines of the Gastern Massif have been broken into several wedges, which are thrust on one another. The lowest of these wedges is seen (Fig. 21) under the word " Rottal glacier," the middle wedge marked R (Rottal hut) is thrust over two smaller wedges, the highest wedge overrides the middle.

The Limestones

A great digitated syncline, delimited by dotted lines in the air, occurs in the limestones, under and to the right of the word " Schwarz Mönch." This syncline belongs to a digitation of the Morcles Nappe, it is called the *involution*.

Ascending from the surface of the foundation of Gastern crystallines towards the lower boundary of the syncline described above (involution), we find four normal series on top of one another (1–4 of the section). The lowest series (1) is the sedimentary cover of the crystallines of the foundation. Series 2, 3 and 4 are slices which have developed owing to the breaking up of the foundation into wedges. These slices belong to the sedimentary cover which has been peeled off the foundation and dragged northwards by the overriding of the granitic mass of the summit of the Jungfrau and Rottalhorn.

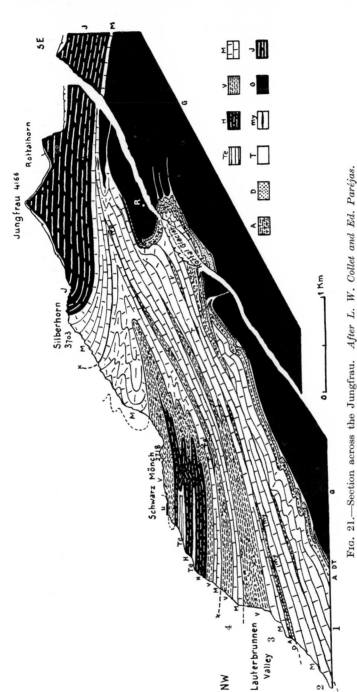

FIG. 21.—Section across the Jungfrau. *After L. W. Collet and Ed. Paréjas.*

Te, Tertiary. *H*, Hauterivian. *V*, Valanginian. *M*, Upper Jurassic. *A*, Argovian. *D*, Dogger (Bajocian and Bathonian). *T*, Trias. *my*, Gneissic mylonites. *G*, Crystallines of the Gastern Massif. *J*, Granite of the Jungfrau (External zone of the Lötschental). *R*, Rottal Hut.

69

On the thrust-plane between series 2 and series 3, near the Lauterbrunnen valley, we meet with two small slices of *gneissic-mylonites*. Following the same thrust-plane towards the Rottal Hut we see above the great crystalline wedge on which the Rottal Hut is built, a cliff of *gneissic mylonites*, marked in black in the section. We shall give later the explanation of

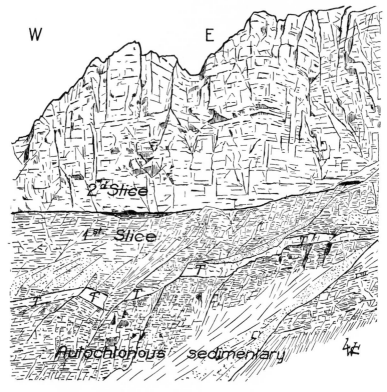

FIG. 22.—On the way to Rottal Hut (Jungfrau).

Autochthonous sedimentary (Trias-Upper Jurassic). 1st slice (*T*, Trias, above : Middle and Upper Jurassic). 2nd slice (Upper Jurassic).

these slices of gneissic mylonites intercalated between two series of sedimentary rocks.

Let us now examine the sedimentary rocks of the Silberhorn, above the upper part of the involution. These sedimentary rocks belong to the sedimentary cover of the granite of the summits of the Jungfrau and Rottalhorn. At the Silberhorn they form an inverted limb.

THE GRANITE OF THE SUMMITS OF THE JUNGFRAU AND ROTTAL- HORN

When following the granite of the summit of the Jungfrau and Rottalhorn towards the west, we see that it represents the *core of the lowest digitation of the Morcles Nappe.* This evidence permits us now to explain the phenomenon of the *involution.*

The Involution. If we compare the sedimentary rocks of the great syncline of the Schwarz Mönch to those of the under-

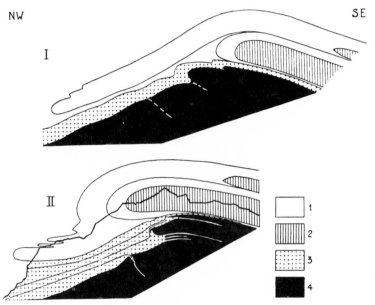

NW   SE

FIG. 23.—The Involution of the Jungfrau.

1, Sedimentary of two digitations of the Morcles nappe. 2, The crystalline core of the two digitations. 3, The sedimentary cover of the Gastern Massif. 4, The crystallines of the Gastern massif.

lying sedimentary slices, we find a great difference. Indeed, the rocks of the slices represent the sedimentary cover of the Gastern Massif, they thus belong to the Autochthon. The rocks of the syncline of the Schwarz Mönch show all the characteristics of the Sedimentary of the Morcles Nappe.

Collet and Paréjas pointed out that the granite of the Jung- frau, with its inverted limb of sedimentary rocks, being the lower digitation of the Morcles Nappe, the digitated syncline of the Schwarz Mönch represent a higher digitation of the Morcles

Nappe which, owing to a phenomenon of involution, has been intercalated between the top of the sedimentary slices of the Autochthon of the Gastern Massif and the lower digitation of the Morcles Nappe, as shown in Fig. 23.

THE JUNGFRAUJOCH AND THE MÖNCH

Up to a height of about 3,600 metres the northern slopes of the Mönch are made up of limestone ; above this point granite of a reddish hue occurs, forming the summit of the mountain. This granite is the same as that forming the summit of the Jungfrau and belongs to the exterior zone of the Lötschental. Erosion has not dissected the slopes of the Mönch to such an extent as those of the Rottal, and the crystalline rocks of the base do not appear.

The carriages of the Jungfrau electric railway convey us to the terminal station, at an altitude of 3,500 metres.

At the station of the Jungfrau a specimen taken from the sides of the tunnel discloses the presence of crystalline rocks of the Gastern Massif. Why are they found at this altitude ? To understand this better, let us pass through the tunnel leading from the restaurant to the platform of the Joch. It has been driven through marmorized limestone in the midst of which a wedge of crystalline rocks—mylonitized gneiss— occurs. Outside the tunnel, in the rocks on which the rain-gauge of the meteorological station stands, we shall find the answer to our question. The upper portion of these rocks is reddish ; it is the granite of the summit of the Jungfrau. These rocks rest on a layer of dark-grey limestone which is none other than the limestone band that we observed at the Rottal dipping into the slopes of the Gletscherhorn. It encloses two wedge-like masses of crystalline rocks.

From the platform of the Joch let us direct our attention to the mighty slopes of the Mönch (Fig. 25). The limestone band supporting the granite of the summit is seen to rise in the direction of the upper Mönchjoch and of the Eigerjoch. Towards the north, in the direction of the spur on which the cabane of the Guggi is built, the limestone band expands to a considerable thickness and forms the slopes dominating the Scheidegg. The inclined rays of the setting sun enable us to distinguish, beneath these dark-grey sedimentary rocks, two crystalline wedges of the Gastern Massif. They are the same

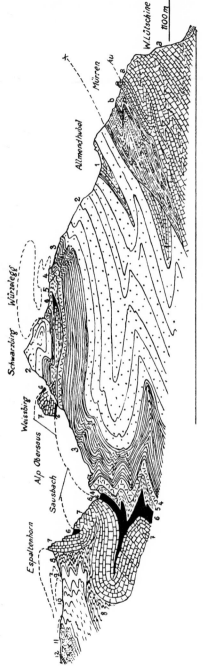

FIG. 24.—Section across the Mürren Region. *After H. Stauffer.*

Autochthon. 4, Upper Jurassic and Lower Cretaceous. *b*, Tertiary. *Wildhorn Nappe.* 1, Aalenian shales. 2, Sandstones (Upper Aalenian and Lower Bajocian). 3, Shales with *Cancellophycus* (Bajocian). 4, Echinodermic limestones (Bajocian). 5, Oxfordian shales. 6, Argovian marls. 7, Upper Jurassic limestones. 8, Berriasian shales. 9, Valanginian limestones. 10, Valanginian shales. 11, Valanginian siliceous limestones (brown hue). 12, Valanginian sandstones and echinodermic limestones.

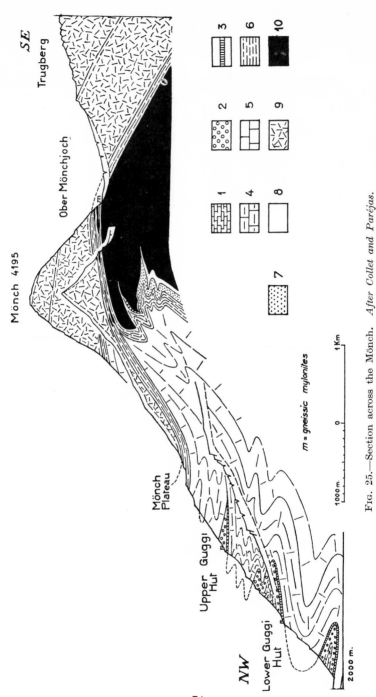

SE

Trugberg

Ober Mönchjoch

Mönch 4195

NW

Lower Guggi
Hut

Upper Guggi
Hut

Mönch
Plateau

m = gneissic mylonites

2000 m.

1000 m.

2000 m.

1 Km

0

1	2	3
4	5	6
7	8	9
		10

FIG. 25.—Section across the Mönch. *After Collet and Paréjas.*

1, Nummulitic limestones and sandstones. 2, Mürren breccia (Priabonian). 3, Eocene, siderolithic. 4, Infravalanginian. 5, Upper Jurassic. 6, Argovian. 7, Middle Jurassic. 8, Trias. 9, Granite of the Jungfrau. 10, Crystallines of the Gastern Massif.

74

PLATE IV. The Rottal Cliff of the
Jungfrau.

Photo by L. W. Collet.

I, Gastern crystallines. *II*, Autochthonous sedimentary. *III*, Sedimentary slices of the Autochthon with gneissic mylonites *M*. *IV*, Morcles Nappe. *V*, Granite (External zone of the Lötschental). *C*, Rottal Hut. *M*, Gneissic mylonites.

two wedges of the Rottal which, thanks to the upward slope of the axis, reappear, as does the limestone, at a higher level. The crystalline rocks which we found at the Jungfrau station belong, therefore, to the uppermost wedge. There is no trace here of the folds we saw beneath the Silberhorn from the Rote Fluh. There is nothing surprising in this. In fact, the northern slopes of the Mönch are much farther to the south than those of the Silberhorn. The crest of one of these folds reappears below the Schneehorn, but, above the cabane of the Guggi, where it should be repeated, it has been removed by erosion.

The ridge, which from the Mönch heads to the south and which bears the Trugberg, is made up of granite of the external zone of the Lötschental dipping beneath the paragneiss of the Lötschental, which in turn passes beneath the Bietschhorn—Aletschhorn—Grimsel granitic Massif.

The Morcles Nappe

The Morcles Nappe was named after the Dents de Morcles, a huge cliff above St. Maurice, on the right bank of the Rhone river (Plate V).

The Stratigraphical Sequence of the Morcles Nappe, in the Dents de Morcles itself, had not yet been published by Lugeon, who is revising this important region. The sequence given in Fig. 26 is based on the studies made by Collet, Paréjas and Moret, between the river Arve and the river Rhone.

Tectonics. The section given in Fig. 28 represents the Morcles Nappe in the Dents du Midi, opposite the Dents de Morcles. The best way to study this interesting nappe is to start from Salvan,[1] on the railway line between Chamonix and Martigny.

The village of Salvan is built on the Carboniferous slates and conglomerates of the southern side of the Aiguilles Rouges Massif. On our way to Salanfe we cross the granite of the Aiguilles Rouges and the crystalline schists containing pink porphyries.

The alluvial plain of Salanfe is an old rock-basin due to

[1] Geological maps : Arpille, 1 : 25,000, by N. Oulianoff, Carte spéciale 103. Francke, Berne. Atlas géologique de la Suisse, 1 : 25,000. Feuille 483 St. Maurice, Francke, Berne, 1934.

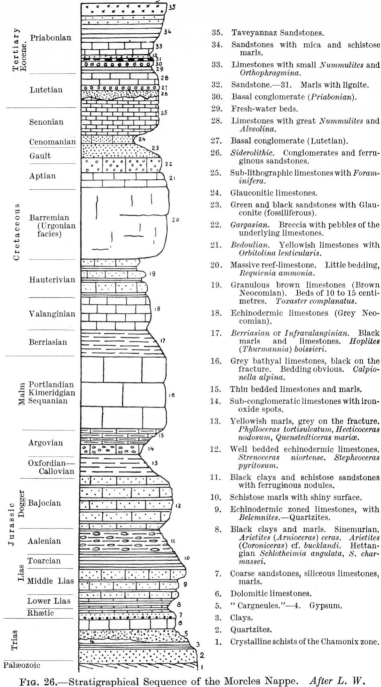

FIG. 26.—Stratigraphical Sequence of the Morcles Nappe. *After L. W. Collet, Ed. Paréjas and Léon Moret.*

76

glacial erosion and filled up by alluvial matter. We have a
good view of the core of the Morcles Nappe in the great wall

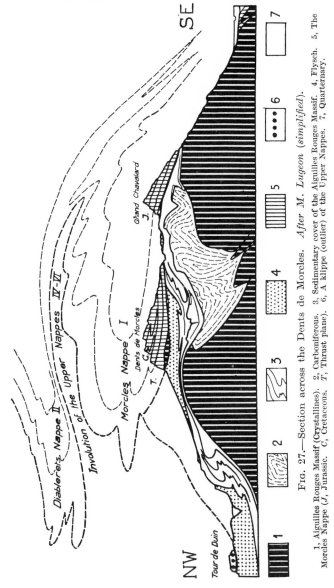

FIG. 27.—Section across the Dents de Morcles. *After M. Lugeon (simplified).*
1, Aiguilles Rouges Massif (Crystallines). 2, Carboniferous. 3, Sedimentary cover of the Aiguilles Rouges Massif. 4, Flysch. 5, The Morcles Nappe (*J*, Jurassic. *C*, Cretaceous. *T*, Thrust plane). 6, A klippe (outlier) of the Upper Nappes. 7, Quaternary.

of the Tour Sallière, where it appears as a digitated recumbent
overfold of Jurassic strata with Cretaceous and Tertiary in

the reversed limb (Fig. 28).   The Dents du Midi represents
the Cretaceous part of this fold, detached along the Infravalanginian and folded separately.

At the Col du Jorat and at the Col d'Emaney the contact of the Morcles Nappe with
the sedimentary cover of the
Aiguilles Rouges is obvious.
The Salentin and the Luisin are
made of gneisses, as is apparent
from the scenery.

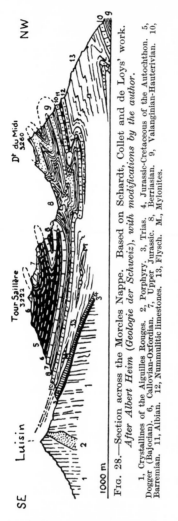

FIG. 28.—Section across the Morcles Nappe. Based on Schardt, Collet and de Loys' work.
After Albert Heim (Geologie der Schweiz), with modifications by the author.
1, Crystallines of the Aiguilles Rouges. 2, Porphyry. 3, Trias. 4, Jurassic-Cretaceous of the Autochthon. 5,
Dogger (Bajocian). 6, Callovian-Oxfordian. 7, Upper Jurassic. 8, Berriasian. 9, Valanginian-Hauterivian. 10,
Barremian. 11, Albian. 12, Nummulitic limestones. 13, Flysch. M., Mylonites.

The ascent of the Col d'Emaney gives one the opportunity
of studying a very interesting
point of the Morcles Nappe, viz.
the existence of a slice of gneissic
mylonites between the sedimentary cover of the Aiguilles
Rouges and the base of the
Morcles Nappe.   On the last
part of the ascent we follow
the Trias which is resting unconformably on the crystalline
schists of the Luisin.   The Col
d'Emaney is due to the weathering of the soft Triassic strata.
On the ridge leading to the
summit of the Luisin we notice
that the Trias is resting on the
old land surface of Permian age.

The mighty slopes of the
Fis—seen from the road to
Chamonix, terminate at the Col
d'Anterne and at Mont Buet,
and are built of Jurassic and
Cretaceous rocks.   These mountains belong to the High Calcareous Alps.   Haug, followed
by Collet, showed several years ago that the limestone mountains of the Fis, together with their foot-hills and the Alps of
Sixt, are geologically connected with the Tour Sallière and the
Dents du Midi.   These mountains have been carved from a

PLATE V.   The Great Cliff of the *Dents de Morcles*, above the Rhone Valley.

Morcles Nappe (above the thrust plane *TP*).   *Nu*, Mummulitic.   *GA*, Gault, Aptian.   *U*, Urgonian.   *B*, Barremian.   *H*, Hauterivian.
ccc, Crystalline slice (mylonites).   *TP*, Thrust plane of the Nappe.   *J*, Upper Jurassic.   *Ar*, Argovian.   *Tr*, Trias.   *t*, Thrust plane in the Autochthon,
Autochthon (below the thrust plane *TP*).—*Fl*, Flysch.   *J*, Upper Jurassic.   *Ar*, Argovian.   *Tr*, Trias.   *t*, Thrust plane in the Autochthon,
marking a duplication.
   Photo by Robert Lugeon.   Geology by Maurice Lugeon.

nappe. If we examine the base of the Fis and the Buet we find the Middle and Upper Lias resting on the Triassic of the Aiguilles Rouges, the separation between them taking place along a *thrust-plane*. Why should the Lias occur here with the same facies as in the Chamonix valley ? It is because we are dealing with a nappe which can be traced from the right bank of the Arve, between Balme-Araches and Sallanches, to the Dents de Morcles on the right bank of the Rhone, and which envelops the folds of the zone of Chamonix.

This is the Morcles Nappe, the lowest of the High Calcareous Alps, the Jurassic core of which is to be seen at Salanfe.

Thus the Morcles Nappe has its roots in the zone of Chamonix.

The Arve Valley.[1] From Sallanches to Cluses the river Arve has cut its way across the Morcles Nappe. A splendid natural section is exposed on the right side of the river Arve, as shown in Fig. 29. Let us study this section from Cluses as far as Le Fayet. At Cluses in front of us the great anticline of the Bargy, formed by Cretaceous strata, is pitching towards the north-east. The river Arve has cut its way across this anticline. At the entrance to the gorge we notice the northern limb of the anticline as seen in the Urgonian strata, then we cross the Hauterivian core and near Balme the southern limb, which is also Urgonian. The reef-limestones of the Urgonian play an important rôle in the topography of the High Calcareous Alps, forming always a white-grey scarp above the Lower Cretaceous ledge.

Two well-defined folds are to be seen after leaving Balme, in the scarp on the right side of the river. The gully above the railway station marks *the mechanical contact of the frontal part of the Morcles Nappe* with the Bargy anticline. The latter belongs to the Autochthon. It has been detached from its base by the northward travel of the Morcles Nappe, and then folded. A few hundred yards to the right of the gully occurs a very good example of a faulted fold, where the Hauterivian comes into contact with the Upper Cretaceous. Four kilometres farther up a recumbent anticline brings the Urgonian higher up where it forms the upper cliff of the mountains (Croix de Fer, Aiguilles de Varens).

The Jurassic core of these folds appears near Oex in the form of a huge S of Upper Jurassic (Malm) with an important

[1] Carte géologique de la France 1 : 80,000. Feuille Annecy.

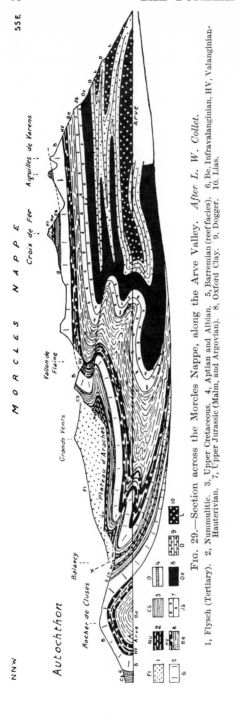

FIG. 29.—Section across the Morcles Nappe, along the Arve Valley. *After L. W. Collet.*

1, Flysch (Tertiary). 2, Nummulitic. 3, Upper Cretaceous. 4, Aptian and Albian. 5, Barremian (reef facies). 6, Be. Infravalanginian, HV, Valanginian-Hauterivian. 7, Upper Jurassic (Malm, and Argovian). 8, Oxford Clay. 9, Dogger. 10. Lias.

cliff of the same rocks thrust over it. The section (Fig. 29) shows that the Jurassic core does not fit into the folds of the Cretaceous cover. We find just the same relation between the folds of the Upper Jurassic and those of the Bajocian and Bathonian (Dogger). As shown by Collet, we have to deal with a disharmonic folding, due to the presence of clay between hard limestones.

From Sallanches we see Mont-Blanc. From this point the Arve valley is no longer transverse, but longitudinal.

Douglas W. Freshfield describes the southern slope of the High Calcareous Alps (Les Fis) as follows:

" The portion of the range which presents its southern face on the traveller's left as he mounts from Sallanches to Servoz may be roughly described as a steep bank from the top of which, but standing back so as to leave room for a level

terrace at its base, rises a turreted wall." At the top of the
latter we recognize the Urgonian reef-limestones, capped by
Upper Cretaceous and Tertiary at its eastern end. The terrace
represents the top part of a huge landslip formed in 1751.

" The breaches in the wall are few, and the tracks up its
face therefore are either, as the Col du Dérochoir, rough
scrambles ; or, as the Echelles de Platé and the sledge-track
which attacks the shoulder of the Aiguille de Varens, the
results of skill in finding the weak point in precipices which,
from many points of view, look vertical, and of labour in building staircases practicable for mountain cattle."

From Le Fayet the large buttress of the Aiguilles Rouges,
the Pormena, is in front of us. To its left a depression eroded
in the Trias marks the contact between the Morcles Nappe and
the sedimentary cover (Autochthon) of the Aiguilles Rouges.

Between the Gemmi Pass and the Aar Valley. Owing
to the pitching, towards the north-east, of the Aiguilles Rouges
Massif, the Morcles Nappe, after having formed the Dents de
Morcles, disappears in the vale of Lizerne. Owing to the change
of pitch, at the north-east end of the transverse depression
situated between the Mont-Blanc Massif and the Aar Massif,
the Morcles Nappe reappears on the Gemmi Pass. The great
corrie of Leukerbad has been carved out of this nappe. The
huge cliff of the Gemmi is made up of Middle and Upper
Jurassic formations of the Morcles Nappe. The Rinderhorn,
Altels, Balmhorn belong to this nappe, the Jurassic core of
which, intensely folded, is beautifully exposed in the lower
part of the Gastern valley. The great cliff running from the
Doldenhorn (southern side) to the Gspaltenhorn, through the
Blümlisalp, is carved out of the Jurassic of the Morcles Nappe.

We have seen previously what happened to the Morcles
Nappe at the Jungfrau (see page 71). From this peak to the
Aar valley we only meet with slices or remnants of the Morcles
Nappe, almost vertical or dipping strongly towards the north,
and covering the tectonic elements of the Autochthon. This
occurs at the foot of the Mettenberg, Wetterhorn and Engel-
hörner, as shown by Scabell and Müller (Fig. 41).

The relation between the frontal part of the nappe and the
roots is shown in Fig. 31. As it is impossible to ascertain
what happens to the Morcles Nappe when it is overlapped by
the Diablerets and Wildhorn Nappes, in the depression between

S.A.

the Mont-Blanc Massif and the Aar Massif, one may ask
whether the nappe dealt with between the Kander valley and
the Aar valley is really the Morcles Nappe ?   The roots of the
nappe studied are situated, as seen in Fig. 31, in the paragneisses of the Lötschental, the equivalent of the Chamonix
zone, where the Morcles Nappe roots.   The correlation then
seems well established, but when we deal with the High
Calcareous Nappes in the Reuss, Linth and Thur valleys, in
the eastern part of Switzerland, matters are not so easy.

Slices of Crystalline Rocks (Mylonites) in the Sedimentary Rocks

When dealing with the geology of the Jungfrau we drew
attention to *crystalline slices* or lenticles (mylonites) which

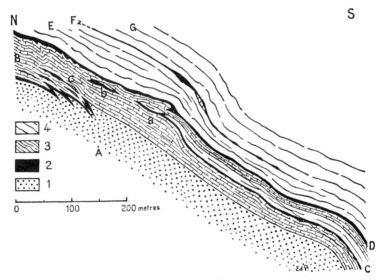

Fig. 30.—The Formation of Lenticles or Slices of Crystalline Mylonites on
the southern side of Jungfrau (Inner-Faflertal, Lötschental). *After
Collet and Paréjas.*

A, Porphyries of the Gastern Massif, showing wedges of which lenticles of mylonites (C)
have been formed.   B, Jurassic limestones in which penetrates a crystalline wedge (C),
with an inverted limb of Trias.   The extremity of this wedge has been broken (a) and a
slice of mylonites (b) has been separated and dragged along in the Jurassic.   D, Inverted
Trias belonging to a slice of paragneisses (E).   G, A mass of crystallines which has been
thrust over (E), and the thrust-plane F is marked by lenticles of Trias and Jurassic.
1, Porphyries. 2, Trias. 3, Jurassic. 4, Paragneisses.

are intercalated in the sedimentary rocks of the cover of the
Gastern Massif.   Moreover, we found also on the thrust-plane

of the Morcles Nappe (p. 78) lenticles of crystalline rocks (mylonites).

How can this type of structure be explained ? Since Collet and Paréjas have studied the southern part of the Jungfrau, near Fafleralp in the Lötschental, the following explanation may be offered, based on field - evidence. These authors have discovered that crystalline lenticles intercalated in sedimentary rocks derive from *wedges* of crystalline rocks. Owing to the laminating and dragging due to higher nappes, the frontal parts of relatively thin wedges of crystalline rocks have been detached from the main wedge. Then they have been carried along intercalated in sedimentary rocks, which formerly were the cover of crystalline rocks, before the formation of the wedges. This peculiar structure is,

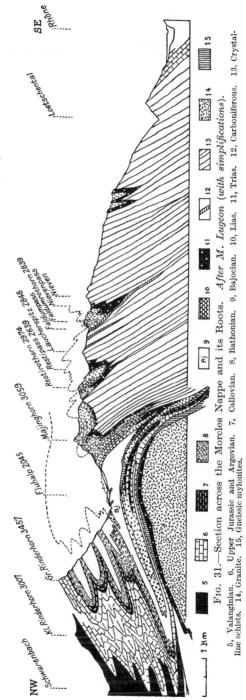

Fig. 31.—Section across the Morcles Nappe and its Roots. *After M. Lugeon (with simplifications).*

5, Valanginian. 6, Upper Jurassic and Argovian. 7, Callovian. 8, Bathonian. 9, Bajocian. 10, Lias. 11, Trias. 12, Carboniferous. 13, Crystalline schists. 14, Granite. 15, Gneissic mylonites.

doubtless, related to *foundation folding* (see p. 17) and is a general feature of the northern side of the Aiguilles Rouges and Gastern Massifs, as shown lately by Collet.

At the Jungfrau, one of these slices of crystalline rocks, intercalated in sedimentary rocks, was discovered a hundred years ago, by Hugi, but as geologists could not find, at the time, any explanation for this observation, it was omitted until rediscovered a few years ago. The same thing occurred for the slice of crystalline rocks on the thrust-plane of the Morcles Nappe.

THE DIABLERETS NAPPE

The Diablerets Nappe consists of a series of sediments beginning with the Middle Jurassic and ending with the Tertiary.

The Diablerets Nappe comes out of the Rhone valley at Ardon. The river debouching at Ardon, the Lizerne, flows for most of its course on the Tertiary (Flysch) of the Morcles Nappe.

The Diablerets Nappe overrides the Morcles Nappe. This phenomenon is well exposed on the left side of the Lizerne valley. Indeed, as shown by Lugeon in his great work, *Les Hautes Alpes Calcaires entre la Lizerne et la Kander*, the Middle Jurassic of the Diablerets Nappe rests on the Tertiary of the Morcles Nappe. *Thus the Diablerets Nappe does not possess a reversed limb* (Fig. 32).

On the way **from Gryon to Anzeindaz** one meets with formations (Trias, Neocomian and Flysch), well exposed at Bovonnaz and farther up at La Tour, near Anzeindaz, which are quite different from those forming the Morcles Nappe and the Diablerets Nappe. Indeed, here we have to deal with the involution of the upper nappes (IV–VI), forming a " cushion," to use Lugeon's word, between the Morcles Nappe and the Diablerets Nappe (see page 64 and Fig. 32).

The Diablerets have been carved out of the frontal part of the nappe, which is enveloped, at the northern base of the mountain, by mighty formations of Tertiary age, containing volcanic ashes, called *Taveyannaz sandstone*.

Owing to the strong pitching to the north-east, the Diablerets Nappe disappears underneath the Sex Rouge and Oldenhorn, which already belong to the Wildhorn Nappe.

That the Diablerets Nappe continues underneath the Wildhorn Nappe is demonstrated by two windows showing Taveyannaz sandstone on the northern border of the High Calcareous Alps, in the vale of Audon and near Gsteig. The last window is due to a transverse fault, of which the eastern side is upthrown, bringing the Taveyannaz sandstones of the Diablerets Nappe to the light.

When **crossing the Sanetsch Pass** the view from the summit towards the Tsanfleuron glacier is very striking. The grey and dull surface of the Nummulitic limestones, supporting the glacier, represents the " carapace " of the Diablerets Nappe, overlain on its northern and southern border by the Wildhorn Nappe.

At the **Gemmi Pass,** where the Morcles Nappe reappears owing to the axis elevation to the north-east, the Diablerets Nappe is also met with. But of the normal series of the Lizerne valley (Middle Jurassic to Tertiary) only a slice of

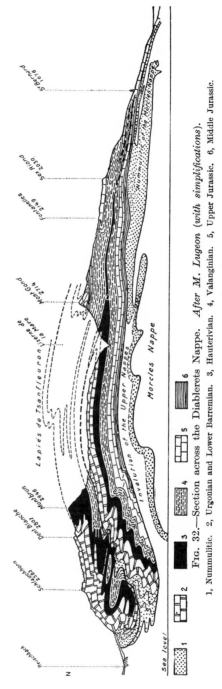

Fig. 32.—Section across the Diablerets Nappe. *After M. Lugeon (with simplifications).*
1, Nummulitic. 2, Urgonian and Lower Barremian. 3, Hauterivian. 4, Valanginian. 5, Upper Jurassic. 6, Middle Jurassic.

Valanginian is exposed, between the Tertiary of the Morcles Nappe and the Middle Jurassic of the Wildhorn Nappe. Under the summit of the Trübelnstock the Valanginian slice is completed by Nummulitic limestones and Taveyannaz sandstones. But farther to the north, on the road leading to Kandersteg, the slice made solely of Neocomian forms the Gellihorn.

Between **Kandersteg** and the **Lauterbrunnen Valley,** as shown by Adrian and Krebs, the slice representing the Diablerets Nappe occurs also. The best exposure is seen at the Birre, on the north-east of the Oeschinen lake. The Diablerets Nappe, indeed, forms the greatest part of the Birre, the summit of which belongs to the Wildhorn Nappe. The slice can be followed towards the north-east as far as the Sefinenfurgge and the Brünli, near Mürren, through the Hohtürli and Dürrenschafberg. At Mürren there is no evidence of the existence of the Morcles Nappe and of the Diablerets Nappe. The Wildhorn Nappe rests directly on the Autochthon (Fig. 24).

Adrian having discovered the presence of *Wildflysch*, at the foot of the Birre, we see that the slice referred to above is the homologue of the Diablerets Nappe. Indeed, the *Wildflysch* plays the rôle of the " cushion " made up of rocks belonging to the upper nappes of the High Calcareous Alps discovered by Lugeon between the Morcles Nappe and the Diablerets Nappe.

To sum up, *the Diablerets Nappe, in the direction of northeasterly axis elevation, has been laminated and reduced to a slice, between the Morcles Nappe below and the Wildhorn Nappe above* (Fig. 33).

THE WILDHORN NAPPE.[1]

In the Lizerne valley, a well-marked recumbent syncline, made up of Nummulitic, Cretaceous and Upper Jurassic, unites the Wildhorn Nappe to the Diablerets Nappe. In the section (Fig. 32) the Wildhorn Nappe appears as a higher digitation of the Diablerets Nappe. But, as already seen, farther to the north-east, at the Gemmi Pass, the Diablerets

[1] Carte géologique des Hautes Alpes Calcaires entre Lizerne et Kander, by M. Lugeon. Carte spéciale 60. 1 : 50,000. Francke, Berne.

Nappe being reduced to a slice of Cretaceous formations, the Wildhorn Nappe occurs as a very independent tectonic element.

The Wildhorn Nappe is the biggest tectonic element of the High Calcareous Alps. It is possible to follow it along the High Calcareous Alps of Switzerland, from the Oldenhorn and Wildhorn as far as the Säntis.

Tectonics. Disharmonic folding is a rule in this nappe. Indeed, the Lower Cretaceous (Valanginian marls) playing the rôle of a lubricant, the Cretaceous and Tertiary formations became detached from their Jurassic core and folded separately. The same phenomenon may occur between the folds of the Upper Jurassic and their Middle Jurassic core, owing to the presence of a lubricant represented by the Oxford Clay. This geological structure is well exposed in the topography, for we generally find several ranges carved out of this nappe. The first range on the south consists of Middle Jurassic strata, the second is made up of Upper Jurassic formations, and the third is carved out of Cretaceous and Tertiary. In several cases the frontal digitations, consisting of Cretaceous and Tertiary formations, may form distinct ranges, as will be seen later on.

In a longitudinal section the Wildhorn Nappe appears as a great saddle (Fig. 34) occurring in the middle of the transversal depression situated between the Mont-Blanc Massif and the Aar Massif. On the highest ridge, Wildhorn-Wildstrubel, outliers of the upper nappes are met with. Fig. 18 shows the relations between the different nappes of the High Calcareous Alps.

The frontal digitations of the Wildhorn Nappe plunge underneath the Prealps, after having been overlain by the upper nappes (IV–VI). This structure is well exemplified at Adelboden, where the Gross Lohner represents the frontal part of the Wildhorn Nappe disappearing underneath the Internal Prealps carved out of the upper nappes (IV–VI). Let us now examine :

The Wildhorn Nappe between the Kander Valley and the Aar Valley.[1] To the east of the Kander valley the summit of the Dündenhorn consists of the Jurassic core of the Wildhorn Nappe, while the Aermighorn and Gerihorn represent the Cretaceous and Tertiary of the frontal part.

[1] Geologische Karte Blümlisalpgruppe by J. Krebs. Carte spéciale **98.** 1 : 25,000. Ibid.

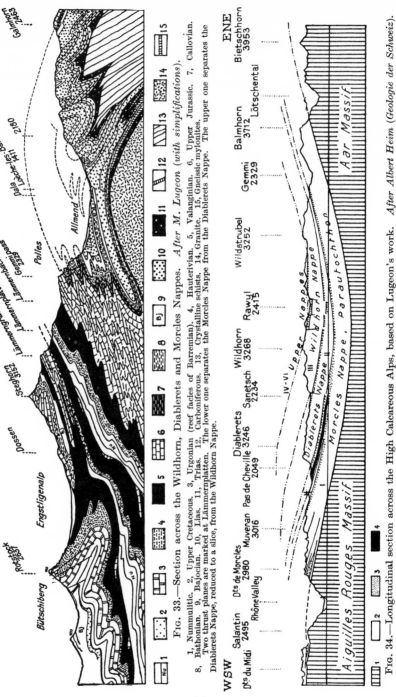

FIG. 33.—Section across the Wildhorn, Diablerets and Morcles Nappes. *After M. Lugeon (with simplifications).*

1, Nummulitic. 2, Upper Cretaceous. 3, Urgonian (reef facies of Barremian). 4, Hauterivian. 5, Valanginian. 6, Upper Jurassic. 7, Callovian. 8, Bathonian. 9, Bajocian. 10, Lias. 11, Trias. 12, Carboniferous. 13, Crystalline schists. 14, Granite. 15, Gneissic mylonites. Two thrust planes are marked at Lämmernplatten. The lower one separates the Morcles Nappe from the Diablerets Nappe. The upper one separates the Diablerets Nappe, reduced to a slice, from the Wildhorn Nappe.

FIG. 34.—Longitudinal section across the High Calcareous Alps, based on Lugeon's work. *After Albert Heim (Geologie der Schweiz).*

1, Hercynian Massifs. 2, Nappes of the High Calcareous Alps. 3, Tertiary. *U*, Involution of Neocomian. 4 and *U* belong to the Upper Nappes. 4, Involution of Trias. *U*, Involution of Neocomian. 4 and *U* belong to the Upper Nappes.

88

Farther to the east the Jurassic core of the nappe forms a very distinct ridge with summits such as the Schilthorn (above Mürren), Männlichen and Faulhorn. At the Schynige Platte we find the frontal folds of the Jurassic core, while on the northern side of the lake of Brienz three ranges were carved out of the frontal part of the Wildhorn Nappe, consisting of Cretaceous and Tertiary. These ranges are : (1) The Harder and Brienzerrothorn, and, farther to the north, (2) the Niederhorn range, and (3) the Sigriswylergrat. An Anticline valley, the Justistal, separates the latter ranges.

Fig. 35 represents a section along the eastern side of the lake of Thun, according to Beck. It shows interesting features on the very front of the Wildhorn Nappe. First of all, we notice between the Harder and the Niederhorn the presence of *Wildflysch*, a Tertiary formation of the upper nappes of the High Calcareous Alps. We shall see later on, when dealing with the External Prealps (see page 237), the mode of formation of these strata, which are well exposed in the banks of the Lombach, the torrent of Habkern. Then we see the Wildhorn Nappe thrust over the Molasse (Tertiary of the Swiss Plateau). But between the thrust plane of the Wildhorn Nappe and the Molasse we find the *subalpine Flysch*, a very complicated formation made up of an interpenetration of slices, of which some belong to the Diablerets Nappe (Taveyannaz sandstone), others to the *Wildflysch* of the upper nappes of the High Calcareous Alps. The presence of *Wildflysch* shows an involution of the upper nappes underneath the frontal part of the Wildhorn Nappe. On the other hand, the slices of Taveyannaz sandstone demonstrate that formations of the Diablerets Nappe have been carried along by the northward travel of the Wildhorn Nappe. We thus have the explanation of the reduction, to a slice, of the Diablerets Nappe in the region of the Gemmi Pass (see Fig. 33).

The Wildhorn Nappe between the Aar Valley and the Reuss Valley.[1] In this region, as shown by Arbenz, the structure of the Wildhorn Nappe is more complicated.

The Jurassic core of the nappe is exposed in the mountains between Engelberg and Frutt, at the Hohenstollen, Hutstock and Widderfeld. It is possible to identify two main digitations

[1] Geologische Karte Engelberg-Meiringen by P. Arbenz. Carte spéciale 55. 1 : 50,000. Ibid.

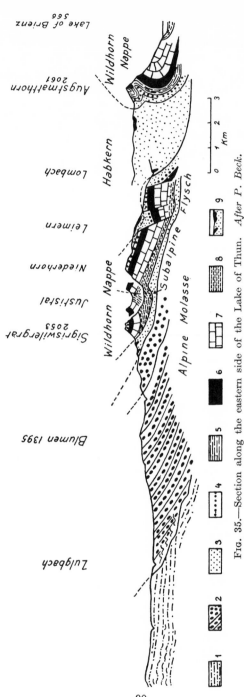

FIG. 35.—Section along the eastern side of the Lake of Thun. *After P. Beck.*

1, Vindobonian. 2, Stampian. 3, Flysch. 4, Eocene. 5, Upper Cretaceous. 6, Urgonian (reef facies of the Barremian). 7, Hauterivian. 8, Valanginian. 9, Wildflysch and Subalpine Flysch.

90

amongst the complications shown by the Jurassic strata (Fig. 36). These digitations, according to Arbenz, are both thrust masses. Having no reversed limb, they represent two distinct t e c t o n i c elements which have been called nappes. *It is the first appearance of the subdivision of the Wildhorn Nappe into two nappes,* which we will study farther to the north-east, later on.

The Cretaceous and Tertiary cover of these digitations were detached from their Jurassic core by the overriding of the Prealp nappes. Indeed, we find remnants, or outliers, of the Prealp nappes in the Tertiary syncline of the frontal part of the Wildhorn Nappe.

These outliers (Klippes) are : the Klippe of the Stanserhorn and the Klippe of the Arvigrat, made up of sediments entirely different from those of the Wildhorn Nappe which they overlap. That we have to deal with outliers is demonstrated by the fact that the Trias of the Klippe of the Stanserhorn rests on the Tertiary of the Wildhorn Nappe.

The Pilatus, which one

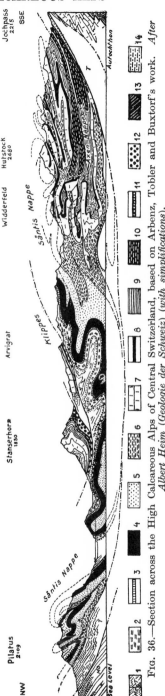

FIG. 36.—Section across the High Calcareous Alps of Central Switzerland, based on Arbenz, Tobler and Buxtorf's work. *After Albert Heim (Geologie der Schweiz) (with simplifications).*

1, Molasse Mo, Flysch and Nummulitic limestones. 2, Upper Cretaceous. 3, Gault. 4, Barremian. 5, Hauterivian. 6, Valanginian. 7, Upper Jurassic. 8, Argovian-Oxfordian. 9, Dogger. 10, Lias. 11, Trias (Autochthon). 12, Trias (Prealps). 13, Crystalline schists. 14, Cretaceous. T, Tertiary.

might easily suppose to be rooting *in situ*, is only the very frontal part of the Wildhorn Nappe thrust over the Molasse of the Swiss Plateau, as shown in Fig. 36.

To sum up, the Wildhorn Nappe shows, in the mountains between the Aar valley and the Reuss valley, two distinct digitations that play the rôle of nappes, viz., a lower element, or the *Axen Nappe*, and an upper one, or the *Säntis Nappe*. This structure will enable us to understand the complications of the Wildhorn Nappe in the north-eastern part of the High Calcareous Alps, viz., between the Reuss valley and the Linth valley.

THE UPPER NAPPES

Introduction. The upper nappes of the High Calcareous Alps are made up of the following elements, in descending order:

VI. The Oberlaubhorn Nappe.

V. The Mont Bonvin Nappe.

IV. The Plaine Morte Nappe.

These nappes come out on the right side of the Rhone valley, between Conthey and Sierre. They cover, as shown on Fig. 17, the southern slope of the Wildhorn Nappe. The terrace passing through the villages of Savièze, Grimisuat, Lens and Montana was carved out of these upper nappes. On the highest ridge, formed by the Wildhorn Nappe between the Wildhorn and the Wildstrubel, the upper nappes have been almost entirely taken away by erosion, with the exception of outliers which cap several summits. On the northern side of the High Calcareous Alps, remnants of these nappes have been spared by erosion.

At the foot of the High Calcareous Alps (northern side) the upper nappes form the *Internal Prealps*, also called the Zone of Passes (Zone des Cols), running from Adelboden to Bex in the Rhone valley. The Internal Prealps, as well as the External Prealps, as will be seen later on, represent the lower element of the Prealps. We thus arrive, with Lugeon, at this important result, that *the External and Internal Prealps are nothing but the upper nappes of the High Calcareous Alps which have been laminated and carried along by the northward travel of the higher nappes of the Prealps, representing the frontal part of the Hinterland.* Thus the region on the right side of the Rhone valley, where we have seen the upper nappes coming out, *contains the roots of the External and Internal Prealps* (Fig. 17).

After this introduction, let us study the relations between the upper nappes and the Wildhorn Nappe. This can be done at Mont Bonvin, a summit rising above Sierre, in the Rhone valley, where interesting observations may be made on the formation of :

The Plaine Morte Nappe. At Mont Bonvin the upper part of the Wildhorn Nappe bends up and forms a recumbent syncline, which is the *joining syncline* between the Wildhorn Nappe and the Plaine Morte Nappe. But of this latter nappe we only know fragments of its reversed limb made up of Cretaceous formations. In rare cases fragments of its normal limb may be met with. These fragments are overridden by the Mont Bonvin Nappe discovered by Schardt.

The following is the sequence of the formations of the Plaine Morte Nappe, according to Lugeon :

Maestrichtian.—Wang formation, consisting of blue bedded limestones with a Serpula: *Jereminella pfenderae* Lugeon.

Turonian.—White limestones, often marly, with Foraminifera.

Barremian.—Passage beds between the bathyal facies and the reef facies, with *Orbitolina conoidea and Gyroporella mühlbergi.*

Hauterivian.—Thin bedded siliceous limestones, of a yellow or brown hue. Often very thick.

Valanginian.—Grey limestones showing a great likeness to the limestones of the Barremian.

Upper Jurassic (Malm).—Compact grey limestones.

These formations show a great likeness to those of the same stages of the Wildhorn Nappe.

The characteristics of the **Mont Bonvin Nappe.** The stratigraphical sequence is quite different from the Plaine Morte Nappe. It was established by Lugeon, as follows :

Flysch (Tertiary).—Coarse sandstones and conglomerates with exotic blocks. Rare Nummulites of the Lutetian. This formation is only met with in the Internal Prealps.

Nummulitic.—Bedded limestones, may be of Priabonian age.

Upper Jurassic (Malm).—Compact limestones.

Argovian.—Rugged limestones.

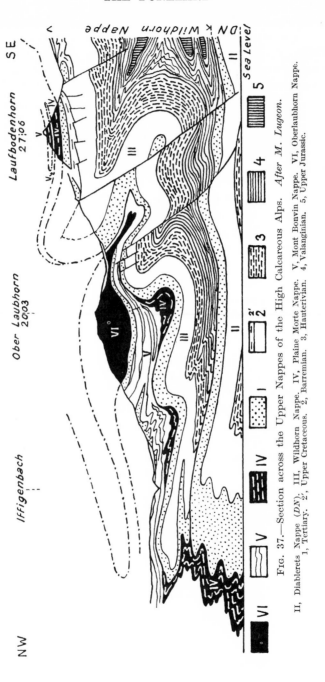

Fig. 37.—Section across the Upper Nappes of the High Calcareous Alps. *After M. Lugeon.*

II, Diablerets Nappe (*DN*). III, Wildhorn Nappe. IV, Plaine Morte Nappe. V, Mont Bonvin Nappe. VI, Oberlaubhorn Nappe.
        1, Tertiary. 2′, Upper Cretaceous. 2, Barremian. 3, Hauterivian. 4, Valanginian. 5, Upper Jurassic.

Callovian-Oxfordian.—Schistose black clays, with Ammonites and iron nodules.

Bathonian.—Echinodermic limestones.

Bajocian.—Shales with mica, of a reddish-brown hue.

Aalenian.—Black shales, representing very likely the whole Upper Lias.

Middle and Lower Lias.—Interstratification of limestones and shales with indeterminable Belemnites. One specimen of *Arietites* has been recorded.

Rhaetic.—Shelly limestones.

Trias.—(1) Gypsum. (2) Dolomitic limestones. (3) Red and green clays.

The Nappe of the Oberlaubhorn consists of Liassic and Triassic formations. It is exposed at the Oberlaubhorn, a summit dividing the Simmen valley into two secondary valleys, to the south of Lenk.

The Tectonic Relations between the Elements of the Upper Nappes. To understand the complicated relations between the upper nappes, let us examine the section shown in Fig. 37. The region through which the section was drawn by Lugeon represents the frontal part of the High Calcareous Alps, to the south of Lenk, in the Simmen valley.

At the summit of the Laufbodenhorn we get the relations between the Wildhorn, Plaine Morte and Mont Bonvin Nappes. Indeed, on the Tertiary of the Wildhorn Nappe we find the Plaine Morte and Mont Bonvin Nappes refolded together. In the syncline of the Wildhorn Nappe, between the Laufboden-horn and Oberlaubhorn, the Plaine Morte Nappe does not exist. Here the Mont Bonvin Nappe rests on the Tertiary of the Wildhorn Nappe.

The Plaine Morte Nappe fills up a syncline of the Wildhorn Nappe, underneath the summit of the Laufbodenhorn, and the same structure is met with farther to the north-west. It is then obvious that the Plaine Morte Nappe has been scratched from the anticlines of the Wildhorn Nappe, by the forward movement of both Mont Bonvin and Oberlaubhorn Nappes.

What is seen of the Oberlaubhorn Nappe is only an outlier or Klippe.

In the Internal Prealps, which represent the most complicated zone in Switzerland, the three elements of the upper nappes of the High Calcareous Alps are reduced to slices, refolded together or interpenetrating each other, owing to the pressure exerted by the overriding of the higher nappes of the Prealps. A detailed mapping of this region has not yet been published, and it is better not to go into details.

The principal outliers of the Plaine Morte and Mont Bonvin Nappes are met with at the Tubang and Mont Bonvin, above Sierre ; at the Rohrbachstein, Laufbodenhorn and Six des Eaux Froides, in the region of the Rawyl Pass ; and last, on the Ammertengrat, a ridge descending to the north from the Wildstrubel. These outliers are well marked in the topography, for they cap summits, the base of which is formed by the Tertiary and Cretaceous of the Wildhorn Nappe.

THE HIGH CALCAREOUS ALPS BETWEEN THE REUSS VALLEY
AND THE RHINE VALLEY.[1]

(Eastern Switzerland)

Introduction. Oberholzer, Albert Heim, Arnold Heim, Arbenz and Buxtorf have correlated the nappes of the High Calcareous Alps situated between the Reuss valley and the Rhine valley with the nappes occurring between the Mont-Blanc Massif and the Aar Massif, as shown in Lugeon's scheme.

The following elements have been ascertained, in descending order :

Säntis Nappe.	⎰Drusberg Nappe. ⎱Räderten Nappe. Säntis Nappe (in re- stricted sense).	
		Wildhorn Nappe.
Axen Nappe.	⎰Toralp Nappe. Silbern Nappe. Bächistock Nappe. Axen Nappe (in re- stricted sense).	

[1] 1. Cartes géologiques spéciales : *Glarner Alpen*, No. 50, by J. Oberholzer and Albert Heim. *Walensee*, No. 44, by Arnold Heim and J. Oberholzer. *Alvier Gruppe*, No. 80, by Arnold Heim and J. Oberholzer. *Säntis*, No. 38, by Albert Heim. *Vierwaldstättersee*, No. 66a, by Buxtorf, Tobler, Niethammer, Arbenz. *Tödi-Vorder Rheinthal*, No. 100A, by F. Weber.

Mürtschen Nappe.⎱ . . . ⎰Homologue of the Diablerets
Glarus Nappe. ⎰   ⎱ Nappe.
Parautochthonous Nappes . Homologue of the Morcles
Nappe.

Tectonics. As already seen, when studying the High Calcareous Alps between the Aar valley and the Reuss valley, the Wildhorn Nappe increases in complication towards the north-east.

Thus the *Axen Nappe* and the *Säntis Nappe* represent two great digitations of the Wildhorn Nappe, though they themselves must be considered as well-defined elements of the High Calcareous Alps. Moreover, the Drusberg and Räderten Nappes, together with the restricted Säntis Nappe, are digitations of the Säntis Nappe ; while the Axen (restricted), Bächistock, Silbern and Toralp Nappes represent digitations of the Axen Nappe.

There is no doubt about the correlation of the Axen and Säntis Nappes with the Wildhorn Nappe, for it is possible to follow on the ground the Wildhorn Nappe along the High Calcareous Alps, from the Wildhorn to the Säntis.

We have seen previously (page 86) that the Diablerets Nappe does not reach Mürren in the Lauterbrunnen valley, where the Wildhorn Nappe rests directly on the Autochthon. Thus, the correlation of the Glarus and Mürtschen Nappes with the Diablerets Nappe may be found unsatisfactory. It is better to say that these nappes are homologous to the Diablerets Nappe, in the sense that they play the same tectonic rôle in another region.

For the same reason, the Parautochthonous nappes may be considered as homologous to the Morcles Nappe.

A glance at the scheme (Fig. 38), drawn by Arnold Heim, enables us to understand the analogies and differences to Lugeon's scheme (Fig. 18). The Aar Massif, though the equivalent of the Mont-Blanc Massif, has a different shape and is broader. The effect on this obstacle of the advance of the Pennine nappes could not be exactly the same as for the Mont-Blanc Massif. Moreover, the northern slope of the Aar Massif was not so steep as those of the Aiguilles Rouges and Gastern Massifs.

A striking feature of Arnold Heim's scheme consists in the existence, at the frontal part of the Säntis Nappe, of a

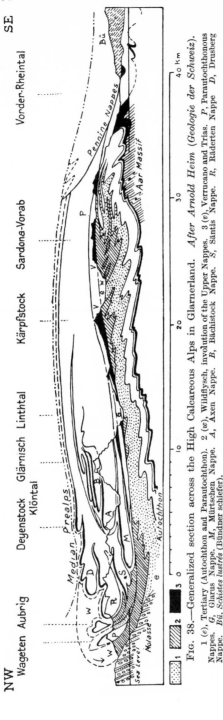

NW

Wageten Aubrig

Deyenstock Glärnisch Linthtal
Klöntal

Kärpfstock

Sardona-Vorab

Vorder-Rheintal

SE

FIG. 38.—Generalized section across the High Calcareous Alps in Glarnerland. *After Arnold Heim (Geologie der Schweiz).*
1 (*e*), Tertiary (Autochthon and Parautochthon). 2 (*w*), Wildflysch, involution of the Upper Nappes. 3 (*v*), Verrucano and Trias. *P*, Parautochthonous Nappes. *G*, Glarus Nappe. *M*, Mürtschen Nappe. *A*, Axen Nappe. *B*, Bächistock Nappe. *S*, Säntis Nappe. *R*, Räderten Nappe *D*, Drusberg Nappe. *Bü, Schistes lustrés* (Bündner schiefer).

patch of mountains, the **Wageten,** which stratigraphically and tectonically represent the frontal part of a Parautochthonous nappe. This structure shows that the Wageten has been dragged away from a Parautochthonous nappe by the travelling of the Wildhorn Nappe.

An involution of the upper nappes underneath the Glarus and Mürtschen Nappes (homologous to the Diablerets Nappe) is obvious. The same phenomenon occurs between the Parautochthonous nappes and the Autochthon. These features reproduce, far to the east, the relations found between the Diablerets Nappe and the Morcles Nappe, and between the Morcles Nappe (Parautochthonous) and the Autochthon.

The homology between the western and eastern High Calcareous Alps is only disturbed by the evidence of a movement of the Wildhorn Nappe

(Axen and Säntis Nappes), after the development of the involution of the upper nappes and the formation of the Parautochthonous nappes.

To sum up we may say that in the eastern part of Switzerland the effect of the Wildhorn Nappe made itself felt on the Parautochthonous nappes, while in the south-western High Calcareous Alps the Diablerets Nappe was reduced and even disappeared.

The Roots of the Nappes.   Standard work has been done by F. Weber on the roots of the High Calcareous Alps in eastern Switzerland.   His very accurate mapping in the north-eastern part of the Aar Massif and in the High Calcareous Alps enables us to understand the relations between the crystalline wedges of the Aar Massif and the nappes of the High Calcareous Alps of this region.   Moreover, the difference in the tectonics of the Aar and Mont-Blanc Massifs will be brought out at the same time.   Let us now examine the section (Fig. 39) drawn by F. Weber, to which I have added the indication of the different zones of the Aar Massif.

Between the Obersandalp and the northern foot of the Bifertenstock, the Gastern-Erstfeld Massif is exposed.   The crystalline substratum of the Bifertenstock consists of para-gneisses which represent the zone of paragneisses of the Lötschental.   The crystalline basis of the Cavestrau is made up of granite belonging to the Central zone of Aar granite.

If we were in the Mont-Blanc Massif we should find the roots of the Parautochthonous nappes in the paragneisses of the Lötschental (Zone of Chamonix).   Weber's section shows that the tectonics of the Autochthon we are accustomed to are exposed on the Gastern-Erstfeld Massif, *as well as on the zone of paragneisses of the Lötschental.*   Indeed, the tectonics of the sedimentary of the Bifertenstock is of the type of the Jungfrau base, though not so complicated.

The roots of the Parautochthonous nappes must then be looked for farther to the south.   Weber has found them in the Central zone of Aar granite, as shown on the section at the Cavestrau and Alp de Schlans.   The very summit of the Cavestrau consists of an involution of the Glarus Nappe, which roots more to the south.

Weber arrives at the very interesting result that the Glarner and Mürtschen Nappes (homologue of the Diablerets Nappe)

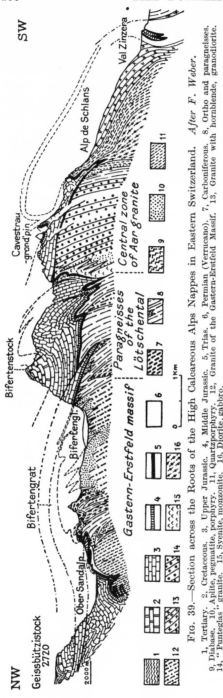

NW

Geissbützistock
2720

Ober Sandalp

Bifertengrat

Bifertenglt.

Bifertenstock

Cavestrau
-grondpin

Alp de Schlans

Val Zinzera

SW

2000 m

0   1 Km

Gastern-Erstfeld massif

Paragneisses
of the
Lötschental

Central zone
of Aar granite

1   2   3   4   5   6   7   8   9   10   11

12   13   14   15   16

FIG. 39.—Section across the Roots of the High Calcareous Alps Nappes in Eastern Switzerland. *After F. Weber.*

1, Tertiary. 2, Cretaceous. 3, Upper Jurassic. 4, Middle Jurassic. 5, Trias. 6, Permian (Verrucano). 7, Carboniferous. 8, Ortho and paragneisses. 9, Diabase. 10, Aplite, pegmatite, porphyry. 11, Quartzporphyry. 12, Granite of the Gastern-Erstfeld Massif. 13, Granite with hornblende, granodiorite. 14, "Punteglas" granite. 15, Syenite, monzonite. 16, Diorite, gabbro.

root in the Tavetsch Massif. The Gothard Massif, on the other hand, operated the Axen and Säntis Nappes (Wildhorn).

These facts show that *the Aar Massif, owing to its greater width, has not been broken up into crystalline wedges in the same crystalline zones as in the Mont-Blanc Massif.*

To sum up, the push of the Pennine nappes produced different effects in the Aar and Mont-Blanc Massifs, so that the nappes of the High Calcareous Alps, which formed in response to the development of the Pennine nappes, could not be identical in the two cases.

The Autochthonous Sedimentary.

The best region to study, from a tectonic point of view, the Autochthonous sedimentary between the Reuss valley and the Rhine valley is the upper part of the Linth valley, above Linthtal. The ascent to the Muttensee Hut of the Swiss Alpine Club, is very interesting, from this point of view. From the Baumgartenalp we cross several slices made up of Cretaceous and Tertiary, while in the mighty cliffs of the Vorder and Mittler Selbsanft, the same structure may be seen in the Upper Jurassic. We have here to deal with *clean-cut thrusts* of the type met with at the base of the Jungfrau. To the north-east of the Kisten Pass, Trias and Middle Jurassic of the Parautochthon rest on the Tertiary of the Autochthon.

The Parautochthonous Nappes. Between the Reuss valley and the Klausen Pass, the Autochthonous fold of the Windgälle is overridden, at the Scheerhorn and Griesstock, by a Parautochthonous nappe which in turn is overlapped by a second Parautochthonous element at the Kammlistock.

The Glarus Mürtschen Nappes. Between the Linth valley, Walensee and Rhine valley, the Autochthon and Parautochthon are overlapped by a huge carapace of continental Permian, called **Verrucano.** The deep valley of the Sernf, in Glarnerland, brings to light the very interesting contact of the travelled mass of *Verrucano* on the Tertiary of the upper nappes (Wildflysch) or directly on the Tertiary of the Parautochthon and Autochthon. Between this Tertiary and the Verrucano, laminated and crushed limestones occur, as recorded by Albert Heim. They generally represent Upper Jurassic formations, but sometimes Triassic, Jurassic and Cretaceous, and they *belong to the reversed limb* of the thrust mass. This reversed limb may be absent and is never thicker than 50 metres.

This **Verrucano carapace** is, in some places, divided into two nappes by the intercalation of Trias. The lower nappe is the *Glarus Nappe*, the upper one the *Mürtschen Nappe.* *Verrucano* does not appear in the core of other nappes of the High Calcareous Alps. So we may say that *Verrucano is the characteristic of the Glarus and Mürtschen Nappes.*

The classical point in order to study the contact of the reversed limb of the Glarner Nappe with its Tertiary substratum is **Lochseite,** on the right side of the Sernf valley, a few kilometres above Schwanden in Glarnerland. The Glarus and Mürtschen Nappes root in the Upper Rhine valley, known as Vorderrheintal. Tectonically and stratigraphically, the *Verrucano* nappes (Glarus and Mürtschen) do not extend so

far as the Klausen Pass to the west, or so far as the Rhine to
the east.

The **crossing of the Segnes Pass** from Flims, in the
Rhine valley, to the Sernf valley in Glarnerland is very interesting.   The village of Flims is built on the upper part of a huge
landslip, which is well marked in the scenery by hills covered
by forests on both sides of the Rhine river, over a distance of
about 14 kilometres and a surface of about 40 square kilometres.   In the depressions between the hills, small lakes are
met with, such as lake Cauma.   The gap to the west of the
Flimserstein, 2·5 kilometres broad, shows the region from
which the landslip originated.   The Flimserstein appears on
our way to the Segnes Pass, as a huge cliff of 500 metres, made
up of Upper Jurassic and Cretaceous formations, belonging to
the normal series of a Parautochthonous fold.   In the northern
part of the Flimserstein *Verrucano*, belonging to the Glarus
Nappe, overrides Cretaceous.   The Segnes Pass (2,625 m.) has
been eroded out of the Glarus Nappe, and the *Flysch* (Tertiary)
of its substratum is exposed there.   To the south-west of the
Pass occurs the great cliff of the Tschingelhörner.   The
different peaks of the ridge are made up of *Verrucano* of darker
hue, resting on a grey cliff that consists of the crushed limestones
of the reversed limb of the Glarus Nappe.   The Flysch of the
Segnes Pass forms the base of the scarp.

The Jurassic and Cretaceous cover of the Glarus and
Mürtschen Nappes are only exposed in the frontal region of
these nappes, in the Linth valley.   The Schilt, a mountain on
the right side of this valley, is made up of the frontal part of
the Glarus Nappe, consisting of Upper Jurassic, Cretaceous
and Tertiary.   On the summit itself, a small cap of Triassic
and Jurassic sediments of the Mürtschen Nappe rests on the
Upper Jurassic of the Glarus Nappe.   The Mürtschen Nappe
forms the Mürtschenstock (2,442 m.) and the Frohnalpstock
has been carved out of the frontal part of this nappe.

The Axen Nappe.   The Axen Nappe is characterized by
the presence of the Lias formation, several hundred metres
thick, and by well-developed Middle Jurassic.   Its southern
margin is represented by the great wall, made up of Jurassic
formations, with Trias at its base, which runs from the
Ortstock, over the Klausen Pass, as far as the Urner lake.
As shown by Albert Heim, the contact between the Axen

Nappe and the Mürtschen Nappe is well exposed on the Klausen Pass. There it is obvious that the Axen Nappe possesses a reversed laminated limb, consisting of crushed Jurassic limestones, resting on the Tertiary of the Mürtschen Nappe. This reversed limb is of the same type as the reversed limb of the Glarus Nappe which we noticed at the Lochseite, in the lower part of the Sernf valley.

The frontal part of the Axen Nappe, consisting of Cretaceous formations, is exposed on the right side of the Urner lake (Reuss valley). There two digitations occur, separated by the Axenmättli Tertiary syncline, which is overturned owing to the plunging of the frontal folds.

The lower *digitations* of the Axen Nappe, called Axen Nappe in restricted sense, and Bächistock Nappe are exposed at the Glärnisch above Glarus. The upper digitations, the Silbern and Toralp Nappes, occur in the Silbern region.

The Säntis Nappe. The Säntis Nappe has travelled 45 kilometres, from the Upper Rhine valley (Vorderrheintal), where it roots, to the Molasse of the Swiss Plateau. The oldest strata of this nappe belong to the Trias, but are only exposed at one place : the Joch Pass, between Engelberg and Meiringen. This case excepted, Lias represents the oldest strata.

The Säntis Nappe, which has been thrust over the Axen Nappe, contains three digitations : the Säntis Nappe (restricted), the Räderten Nappe and the Drusberg Nappe. Owing to southward pitch, the lower digitations (Säntis Nappe (restricted) and Räderten Nappe) disappear underneath the upper one, the Drusberg Nappe. It is possible to follow the Säntis Nappe from Drusberg to the Wildhorn Nappe, through Frohnalpstock, Brünig, Brienzerrothorn, Harder, Morgenberghorn, Dreispitz, viz., over a distance of 110 kilometres.

The relations between the Säntis Nappe and the Axen Nappe are shown in Fig. 40. To the north-west of the Axen Nappe, already studied in this section, occurs an overturned syncline of Tertiary, the *Riemenstalden syncline*, discovered by Arbenz, which separates the Axen Nappe from the Säntis Nappe.

The Frohnalpstock (Fig. 40), which commands the villages of Brunnen and Morschach, has been carved out of the Säntis Nappe. The Rigihochfluh, a calcareous range situated between the Muota river and the Rigi, represents a normal series of

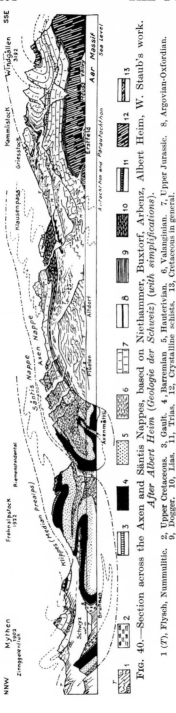

NNW

Mythen 1902
Zinggelenfluh

Frohnalpstock 1932

Riemenstaldental

Klippes (Median prealps)

Schwyz

Brunnen

Axenmättli

Flüelen

Altdorf

Fellis Valley

Aar Massif

Sea level

Autochthon and Paraautochthon

Säntis Nappe

Axen Nappe

Klausenpass

Griesstock

Kammlistock

Windgällen 3192

SSE

Ersfeld

FIG. 40.—Section across the Axen and Säntis Nappes, based on Niethammer, Buxtorf, Arbenz, Albert Heim, W. Staub's work. After Albert Heim (*Geologie der Schweiz*) (*with simplifications*).

1 (*T*), Flysch, Nummulitic. 2, Upper Cretaceous. 3, Gault. 4, Barremian 5, Hauterivian. 6, Valanginian. 7, Upper Jurassic. 8, Argovian-Oxfordian. 9, Dogger. 10, Lias. 11, Trias. 12, Crystalline schists. 13, Cretaceous in general.

Cretaceous and Tertiary uniting with the anticline Brunnen-Morschach, by means of a Tertiary syncline (Fig. 40).

At Schwyz the latter Tertiary syncline widens and the Mythen rest on it, giving the impression that they are "swimming in Flysch." Indeed, the Mythen (Plate X), the mountains above Schwyz, are an outlier (klippe) of the higher nappes of the Prealps. The oldest strata of the Mythen, the Trias, rest on the *Flysch* (Tertiary) of the High Calcareous Alps. *The Mythen represent an outlier of the Median Prealps* (see p. 249).

To the east of the Linth valley, the mighty cliff of the Churfirsten, on the northern side of the Walen lake, is made up of three different tectonic elements in ascending order: (1) The Mürtschen Nappe; (2) a slice of the Axen Nappe; and (3) the Säntis Nappe.

The slice of the Axen Nappe is what has been called by Arnold Heim the "Walenstadt Zwischendecke." This slice shows that the Axen Nappe dies away to the north-east. In the northern part of the Churfirsten the Säntis Nappe directly overrides the Mürtschen Nappe.

The Säntis represents the geological continuation of the Churfirsten, to the north. This mountain has been carved out of the frontal part of the Sänthis Nappe. As shown by Albert Heim, six anticlines are met with, made up only of Cretaceous formations. The scenery of the Säntis, with its very steep calcareous slopes, is due to the fact that both rain and snow water percolate through the limestones. In fact, no river is met with in this region.

BIBLIOGRAPHY

A. HIGH CALCAREOUS ALPS (SWITZERLAND)

1. HEIM, ALB.—Das Säntisgebirge. Beitr. zur geol. Karte d. Schweiz. N.F. 16e Lief. (46). Bern, 1905.

2. BUXTORF, A. TRUNINGER, E. —Ueber die Geologie der Doldenhorn-Fisistock-gruppe und den Gebirgsbau am Westende des Aarmassivs. Verhandl. d. Naturf. Ges. in Basel, Bd. xx, H. 2, 1909.

3. HEIM, ARN.—Monographie der Churfirsten-Mattstock Gruppe. Beitr. zur geol. Karte d. Schweiz. N.F. 20e Lief. (50). Bern, 1910.

4. COLLET, L. W.—Les Hautes Alpes calcaires entre Arve et Rhône. Mém. Soc. Phys. et Hist. nat. Genève, vol. 36, fasc. 4, 4°, fig. pl. carte, 1910.

5. BECK, P.—Geologie der Gebirge nördlich von Interlaken. Beitr. zur geol. Karte d. Schweiz. N.F. 29e Lief. Bern, 1910.

6. STAUB, W.—Geologische Beschreibung der Gebirge zwischen Schächental und Maderanertal im Kanton Uri. Beitr. z. geol. Karte d. Schweiz. N.F. 32e Lief. (62). Bern, 1911.

7. BECK, P.—Beiträge zur Geologie der Thunerseegebirge. Beitr. zur geol. Karte d. Schweiz, N.F. 29e Lief. Bern, 1911.

8. ARBENZ, P.—Die Faltenbogen der Zentral- u. Ostschweiz. Viertelj. Schrift der Naturforsch. Gesell. in Zürich. LVIII, 1913.

9. ADRIAN, H.—Geologische Untersuchung der beiden Seiten des Kandertals in Berner Oberland. Eclogae geol. Helvet., XIII, p. 238, 1915.

10. LUGEON, M.—Les Hautes Alpes calcaires entre la Lizerne et la Kander. Mat. carte geol. Suisse, N.S., Livr. 30. Berne, 1914–1918.

11. HEIM, ARN.—Zur Geologie des Grünten im Allgäu. Festschrift Albert Heim, Vierteljahrsschrift der Naturf. Gesellsch. Zürich, 1919.

12. STAUFFER, H.—Geologische Untersuchung der Schilthorngruppe im Berner Oberland. Mitteil. d. Naturf. Ges. Bern. Jahrg. 1920, H. 1.

13. Paréjas, Ed.—Géologie de la Zone de Chamonix. Mém. Soc. Phys. Hist. nat. Genève, vol. 39, fasc. 7, 1922.

14. Heim, Arn.—Beobachtungen in den Vorarlberger Kreideketten. Eclogae geol. Helvet., vol. xviii, N° 2, 1923, p. 207.

15. Heim, Arn.—Der Alpenrand zwischen Appenzell u. Rheintal (Fähnern Gruppe) und das Problem der Kreide Nummuliten. Beitr. z. geol. Karte d. Schweiz, N.F. 53e Lief. Bern, 1923.

16. Louis, K.—Beiträge zur Geologie der Männlichen-Gruppe (Berner Oberland). Jahrb. d. Phil. Fakult. ii, Univ. Bern, Bd. v, 8°, nov. 1924.

17. Thiel, Piet van.—Geologische Forschungen zwischen Bezau und Egg (Vorarlberg). 8°, La Haye, 1924.

18. Günzler-Seiffert, H.—Der geologische Bau der östlichen Faulhorngruppe im Berner Oberland. Eclogae geol. Helvet., vol. xix, N° 1, Bâle, 1924.

19. Buxtorf, A.—Geologie des Pilatus. 105e Jahresversammlung d. Schw. Naturf. Ges. Luzern 1924, 8°, 17 pp., fig. pl. Bern, 1924.

20. Krebs, J.—Geologische Beschreibung der Blümlisalpgruppe. Beitr. z. geol. Karte d. Schweiz N.F. 54e Lief. (iii). Bern, 1925.

21. Sax, H. G. J.—Geologische Untersuchungen zwischen Bregenzer Ach und Hohem Freschen (Vorarlberg). In. Diss. Zürich, 8°, 1925.

22. Schaad, H. W.—Geologische Untersuchungen in der südlichen Vorarlberger Kreide-Flyschzone zwischen Feldkirch und Hochfreschen (Deutschösterreich). In. Diss. Zürich, 4°, 1925.

23. Gagnebin, E.—Une lame de gneiss parautochtone à la base de la Dent du Midi (Ecaille du Jorat). Bull. Soc. vaud. Sc. nat., vol. 56, 1925, N° 216.

24. Schaad, H. W.—Zur Geologie der jurassischen Kanisfluh- Mittagfluhgruppe im Bregenzerwald. Vierteljahrsschrift der Naturforsch. Gesellsch. Zürich, 71, Jahrg. 1926, 1. und 2. Heft, pp. 49–84.

25. Schaad, H. W.—Beiträge zur Valangien- und Hauterivien Stratigraphie in Vorarlberg. Geol. Rundschau xvii, p. 81, 1926.

26. Lugeon, M.—Quelques lignes sur les nappes des Diablerets et du Wildhorn (Alpes de la Suisse occidentale). Livre jubilaire 50re Fondat. Soc. géol. de Belgique, 4°, Liège, 1926, pp. 359–361.

27. Meesmann, Paul.—Geologische Untersuchung der Kreideketten des Alpenrandes im Gebiet des Bodenseerheintals. Thèse, Bâle, 1925, 111 pp.

28. Bonnard, Emile G.—Monographie géologique du massif du Haut de Cry Beiträge z. Geol. k. der Schweiz, N.F. 57 Lief. iv Abt. Bern, 1926, 57 pp.

29. Goldschmid, K.—Geologie der Morgenberghorn-Schwalmer-gruppe bei Interlaken. Mitt. Naturf. Gesell. Bern, 1926. Bern, 1927, 73 pp.

30. Luther, M.—Die tektonischen und stratigraphischen Zusammenhänge östlich und westlich der Reuss zwischen Brunnen und Amsteg. Jahrb. d. Phil. Fakult. ii, Univ. Bern, Bd. vii, 1927.

31. Loys, François de.—Monographie géologique de la Dent du Midi. Annexe (Pl. ii): La chaîne de la Dent du Midi aux Dents Blanches de Champéry, vue de la Croix de Culet, panorama géol. par Elie Gagnebin. Mat. Carte géol. Suisse, N.S., 58e liv., 1928.

32. SCHUMACHER, P. V.—Der geologische Bau der Claridenhette. Mat. Carte geol. Suisse, N.S., 50e liv., ive partie, 1928.
33. LIECHTI, P.—Geologische Untersuchung der Dreispitz-Standfluh-gruppe und der Flyschregion südlich des Thunersees. Thèse Mitt. Naturf. Gesell. Bern (1930), pp. 77–206.
34. LUGEON, MAURICE.—Trois tempêtes orogéniques à la Dent du Midi. Livre Jubilaire Soc. Géol. France, 1930, ii, pp. 499–512.
35. COLLET, L. W., et PARÉJAS, ED. { Géologie de la Chaîne de la Jungfrau. Mat. Carte Géol. Suisse. Nouvelle série. 63e liv., 1931.
36. OBERHOLZER, J.—Geologie der Glarneralpen. Mat. carte géol., Suisse, N.S., 28ème liv., 1933.

B. HIGH CALCAREOUS ALPS OF FRANCE OR CHAINES SUBALPINES

1. KILIAN, W.—Description géologique de la Montagne de Lure (Basses-Alpes). Paris, 1889, 8°, fig. pl. c.
2. HAUG, E.—Les chaînes subalpines entre Gap et Digne. Bull. Serv. Carte géol. France. T. iii, N° 21. Paris, 1891–92.
3. LUGEON, M.—Les dislocations des Bauges (Savoie). Bull. Serv. Carte géol. France, N° 77, T. xi, 1899–1900.
4. RÉUNION EXTRAORDINAIRE de la Société géologique de France dans les Alpes Maritimes 9–19 septembre 1902. Excursions préliminaires du 5 au 8. Bull. Soc. géol. France, 4ème s. T. 2, p. 509, 1902.
5. KILIAN, W., et RÉVIL, J. { Etudes géologiques dans les Alpes occidentales. Mém. Carte géol. dét. France, 2 vol., 4°. Paris, 1904–1917.
6. COLLET, L. W.—Les Hautes Alpes calcaires entre Arve et Rhône. Mém. Soc. Phys. Hist. nat. Genève, vol. 36, fasc. 4, 1910.
7. RÉVIL, J.—Géologie des chaînes jurassiennes et subalpines de la Savoie, 2 vol., 8°. Chambéry, 1911–1913.
8. BOUSSAC, J.—Etudes stratigraphiques sur le Nummulitique alpin. Mém. Carte géol. dét. France, 4°, Paris, 1912.
9. KILIAN, W.—Aperçu sommaire de la géologie, de l'orographie et de l'hydrographie des Alpes dauphinoises. 2ème édition. Annales Univ. Grenoble, T. xxxi, N° 1, 1919.
10. LES RÉGIONS jurassienne, subalpine et alpine de la Savoie. Réunion extraordinaire de la Société géologique de France du 14 au 20 septembre 1921. Paris, Soc. géol. de Fr. 1922.
11. KILIAN, W.—Sur la structure des chaînes subalpines dauphinoises et du Vercors méridional. Annales Univ. Grenoble, T. i, N° 1 (nouvelle série), Grenoble, 1924.
12. RÉUNION EXTRAORDINAIRE de la Société géologique de France dans le Gard, le Vaucluse et la Drôme du 10 au 18 septembre 1923. Bull. Soc. géol. France, 4ème s. T. 23, p. 461, 1924.
13. KILIAN, W., et LANQUINE, A. { Sur la tectonique des chaînons les plus externes des Alpes entre Chabrières et Moustier-Sainte-Marie (Basses-Alpes) et sur les faciès des terrains qui les constituent. Trav. Lab. géol. Univ. Grenoble 1924, T. xiii, fasc. 2, p. 117.

14. HAUG, E.—Les nappes de charriage de la Basse Provence.   Mém.
    Carte géol. dét. France.   Paris, 4º, 1925.
15. PARÉJAS, ED.—La tectonique du Mont Joly (Hte Savoie).   Eclogae
    geol. Helvet. XIX, Nº 2, pp. 420–503, fig. pl. 1925.
16. MORET, L.—Monographie géologique du Roc de Chère (Lac d'An-
    necy). Bull. Serv. Carte géol. France, Nº 159, T. XXIX,
    1925–26.
17. NASH, J. -M. W.—De Geologie der Grande Chartreuse Ketens.
    Thèse Grenoble, 8º, Delft, 1926.
18. PARÉJAS, ED.—Nouvelles observations sur le soubassement du
    Mont Joly (Haute-Savoie).   Eclogae geol. Helvet. T. XX, Nº 2,
    p. 331.   1927.
19. SCHOELLER, H.—La nappe de l'Embrunais au Nord de l'Isère.
    Bull. Carte géol. France, Nº 175.   Paris, 1929.
20. GIGNOUX, M.,⌠Les grandes subdivisions géologiques des Alpes
    et MORET, L.⌡françaises.   Annales de Géographie, 15 juillet 1934.
21. MORET, L.—Géologie du Massif des Bornes.   Mém. Soc. Géol. de
    France, Nº 22.   1934.
22. BLANCHET, F.—Etude géologique des montagnes d'Escreins.
    Allier, Grenoble, 1934.

CHAPTER VI

THE GEOLOGY OF THE GRINDELWALD REGION

THE DIFFERENT TECTONIC ELEMENTS

The different tectonic elements of the Grindelwald region are the following :

1. The Aar Massif.
2. The High Calcareous Alps.

The Aar Massif. Several zones of the Aar Massif are represented to the south of Grindelwald, viz. : (*a*) the Gastern Massif ; (*b*) the Erstfeld Massif ; (*c*) the Central zone of the Aar granite.

The principal summits belonging to the Gastern Massif are : the Mittelhorn (Wetterhorn, 3,708 m.) and the Mettenberg (3,107 m.).

The Erstfeld Massif includes the Gross Fiescherhorn (4,049 m.), Klein Fiescherhorn (3,905 m.) and the Gross Schreckhorn (4,090 m.).

The Finsteraarhorn has been carved out of the Central zone of the Aar granite.

The High Calcareous Alps. In the Grindelwald region the Calcareous Alps are formed by three different tectonic elements. They are :

(*a*) The Parautochthonous Nappe (Morcles), made up of several slices, which may be seen at the base of the cliff of the Wetterhorn and of the Mettenberg, then also at Schüssellauenen, to the south of Grindelwald.

(*b*) The Wildhorn Nappe, with the following summits : the Schwarzhorn (2,930 m.), Faulhorn (2,684 m.), Männlichen (2,346 m.).

(*c*) The upper nappes of the High Calcareous Alps (IV–VI), to which probably belong the Galtbachhorn (2,319 m.), Lauberhorn (2,475 m.) and Tschuggen (2,523 m.).

Stratigraphy. Let us study these different elements (Mesozoic and Tertiary) from a stratigraphical point of view.

The Autochthonous Mesozoic rests unconformably on the crystalline schists of the Gastern Massif. It consists of the following formations, from bottom to top : Trias, Dogger (Bajocian-Bathonian), Argovian, Upper Jurassic (Malm), Infravalanginian, Nummulitic. The Trias is sometimes lacking, and then the Bajocian transgresses on the crystalline of Gastern. During the Tertiary the Alpine movements broke up this massif into enormous wedges, the space between them being filled up by crushed sedimentary rocks.

The Wetterhorn (3,703 m.) and Eiger (3,974 m.) have been carved out of the Upper Jurassic (Malm) of the Autochthon.

Slices belonging to the Parautochthonous nappe (Morcles) are locally found lying on the Autochthon. They are made up of Upper Jurassic, Lower Cretaceous and Tertiary.

The Wildhorn Nappe rests on the Parautochthon and consists, in the neighbourhood of Grindelwald, of Upper Lias (Aalenian), Dogger, Callovian, Oxfordian and Argovian.

Elements probably belonging to the upper nappes (IV–VI) of the High Calcareous Alps are represented by the Aalenian slice of Tschuggen and the shales of the same age occurring at the Kleine Scheidegg, as shown by Arbenz.

The relations existing between these different elements will stand out more clearly, if we follow them along certain itineraries.

FROM GRINDELWALD
TO THE BÄREGG AND THE SCHWARZEGG HUT OF THE SWISS ALPINE CLUB

The path crosses the Schwarze Lütschine and then passes near the quarry of the so-called " Grindelwald marbles." This formation represents the Infravalanginian of the Autochthon, with many impregnations of Tertiary Siderolithic (ironstone). Higher up, we pass from the Infravalanginian to the Upper Jurassic (Malm) ; the Bäregg Inn is built on the last formation. Farther, the track passes across the Argovian, the Callovian and Bathonian ferruginous Oolites and the echinodermic limestones of the Bajocian.

At Ortfluh, the Trias refolded is met with, resting on the crystalline substratum.

Proceeding upwards, from the Bäregg to the Mettenberg, we notice that this series of the Autochthon forms a great recum-

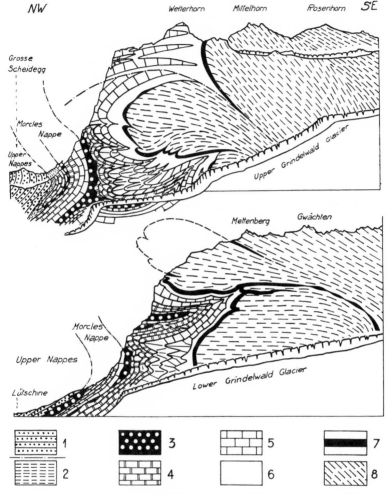

FIG. 41.—Sections across the Wetterhorn and Mettenberg, near Grindelwald. *After Scabell, F. Müller and Ed. Paréjas.*

1, Middle Jurassic of the Upper Nappes, High Calcareous Alps. 2, Flysch. 3, Nummulitic breccias (Priabonian). 4, Lower Cretaceous. 5, Upper Jurassic. 6, Argovian. 7, Trias-Middle Jurassic. 8, Crystallines.

bent syncline. Its Tertiary core may be reached at Hohturnen (Fig. 41). This syncline is overridden by the crystalline wedge, out of which the summit of the Mettenberg is carved.

In the Upper Jurassic of the left bank of the Lower Grindelwald glacier, cleavage is particularly apparent, crossing bedding-planes which show small folds.

Having passed the Bäregg, we walk along the lower element of the crystallines. It stretches as far as Rotgufer (Fig. 41). There sedimentaries appear above the path, very strongly crushed and separate the crystalline base from the crystalline wedge of the summit of the Mettenberg.

The Klein Schreckhorn belongs to an even more internal crystalline wedge. The Schwarzegg Hut of the Swiss Alpine Club is built on this tectonic element. Its sedimentary cover may be seen on the topmost ridge running along the northwest ridge of the Klein Schreckhorn. It does not, however, reach the path we are following.

From the Bäregg

to the Bergli Hut of the Swiss Alpine Club and the Jungfraujoch [1]

The glacier has to be crossed in order to reach the Bergli Hut from the Bäregg. Then the way goes through the Kalli, which is made up of the lower crystalline element we already crossed between the Bäregg and Rotgufer, on the right bank of the lower Grindelwald glacier.

The unconformity of the Trias on the crystalline schists may be observed on reaching the Kalliband. Above the Trias lie successively the Dogger (echinodermic limestones) and the Argovian (spotted limestones with ankerite, and schistose limestones). The greatest part of the long cliff of the Kalliband consists of this last formation. Farther to the north, above the Kallifirn, rises the mighty Jurassic wall of the Eiger, showing a syncline bend above the railway station of Eismeer.

The Bergli rocks, on which is built the Swiss Alpine Club Hut, also belongs to the crystallines of Gastern. Higher up, at the Unter Mönchjoch and at the Ober Mönchjoch, the Jurassic limestones of the upper band of sedimentary of the Jungfrau are met with in rocky islands showing through the surface of the glacier. To the limestones, indeed, are due the passes as well as the gently slanting surface covered up

[1] Carte géologique spéciale, *Jungfrau*, No. 113, by Collet and Paréjas.

by the glacier, and over which we reach the Jungfraujoch without effort. (For the Geology of the Jungfraujoch, see page 72.)

FROM GRINDELWALD

TO THE KLEINE SCHEIDEGG

The path leading from Grindelwald to the Kleine Scheidegg follows the foot of the north-western cliff of the Eiger, which is formed by complicated folds of the Autochthon, consisting of Upper Jurassic, Infravalanginian and Tertiary.

In the lower part of the Schüssellauenen gully, to the south of Grindelwald, three Parautochthonous slices (Morcles Nappe) are seen resting on the Autochthon of the Eiger. The Lower Cretaceous and the Tertiary only participate in the formation of these slices.

The railway line, from Grindelwald to the Kleine Scheidegg, is mostly cut out of moraines, as far as Alpiglen. From this station upwards, it very nearly follows the contact between the Nummulitic limestones and the black shales of the *Flysch* (Tertiary).

The frontal folds of the Eiger, made up of Tertiary formations (Siderolithic and Nummulitic), are well exposed along the Krutwaldbach, above Alpiglen.

The whole district lying between Grindelwald, Wengen and Zweilütschinen is covered by the Wildhorn Nappe, which has overridden the Parautochthon and the Autochthon.

The skeleton of this region is formed by the Upper Lias, Dogger, Callovian, Oxfordian and Argovian strongly folded.

The Kleine Scheidegg, as well as the summits of the Lauberhorn, Tschuggen and Galtbachhorn, have been carved out of Aalenian slices, which, according to Arbenz, may belong to the upper nappes of the High Calcareous Alps (IV–VI).

FROM GRINDELWALD

TO THE GROSSE SCHEIDEGG

On the way from Grindelwald to the Grosse Scheidegg, the profile of the Mettenberg, previously described when climbing to the Schwarzegg Hut, is completed by the profile of the Wetterhorn (Fig. 41). The crystalline wedge of the Mettenberg

S.A.

summit expands to the north-east and appears entirely embedded in the limestones of the Upper Jurassic (Malm). The latter forms the greatest part of the slopes between the Klein Wellhorn and the Mettenberg.

At the foot of the slopes of the last mountains, on the Autochthon, the slices of the Parautochthonous nappe overlie each other. There may be as many as three of them. The rocks forming these slices belong to the Upper Jurassic (Malm), Infravalanginian, Hauterivian and Tertiary (quartzites, limestones, Taveyannaz sandstones and Flysch). One of these slices can easily be studied on the left bank of the Schwarze Lütschine above " Auf der Sulz." There, the sandstones and echinodermic limestones of the Hauterivian rest on the light grey limestones of the Infravalanginian.

During the whole ascent, from Grindelwald to the Grosse Scheidegg, we walk on the Upper Lias (Aalenian) of the Wildhorn Nappe, in which the *Ludwigia murchisonœ* and *Lioceras opalinum* zones are represented.

At the Grosse Scheidegg itself the Wildhorn Nappe is seen overlying the slices of the Parautochthon which are upraised and even overturned.

The whole Faulhorn and Schwarzhorn region belongs to the Wildhorn Nappe.

CHAPTER VII

THE SWISS PLATEAU

INTRODUCTION

The Swiss Plateau is situated between the Alps and the Jura Range. It is a syncline joining the Jura to the Alps, filled up by Tertiary formations. From a tectonic point of view we may say that the Swiss Plateau is nothing else than a syncline between the most external elements (the Jura being taken as one element) of the virgation of the Western Alps. It begins at Aix les Bains, where the Jura separates from the Alps, and widens rapidly to the north-east.

Swiss geologists generally call this region the "*Plateau Molassique*," for the Tertiary formations of Upper Oligocene and Miocene age are almost entirely made up of soft sandstones of fresh-water or marine origin called "Molasse." The marine Molasse is used as building stone, but gives, owing to its grey hue, a dull aspect to many Swiss towns. The alluvial matter brought down by rivers and streams escaping from the recently formed Alps represents the constituents of the Molasse. The coarser deposits thus correspond to the shore line at the foot of the Alps. One can see true deltas deposited under fresh-water and under marine conditions. The conglomerates which characterize these deltas have been called "Nagelfluh." They play an important rôle in the topography, forming great cliffs as at the Righi, or important reliefs as at the Napf (1,408 m.).

STRATIGRAPHY OF THE MOLASSE

The following is the sequence of the principal beds of the Molasse, from top to bottom :

4. The Upper fresh-water Molasse. *Tortonian* (Upper Miocene).

3. The Upper marine Molasse. *Burdigalian and Helvetian* (Lower and Middle Miocene).

115

2. The Lower fresh-water Molasse. *Chattian-Aquitanian* (Upper Middle Oligocene to Upper Oligocene).

1. The Lower marine Molasse. *Rupelian* (Lower Middle Oligocene).

This succession of beds shows at the first glance that we have to deal with two continental periods separated by a marine transgression. To explain these facts we shall have to consider, later on, contemporaneous movements that took place in the Alps.

1. **Lower Marine Molasse.** Sandstones representing a marine transgression on continental formations. These marine beds occur on the northern border of the Autochthonous chains between the river Arve and Annecy.

2. **Lower fresh-water Molasse.** The lower beds are generally of a reddish colour due to the iron oxides brought down to the big lake by the streams. This colour shows that the climate on the land was much warmer than at present, and more humid. The upper beds are grey. The lower fresh-water Molasse contains *lignites* due to the deposit of plant remains transported from the land by the streams. We find also fresh-water limestones, beds of gypsum and clays. The gypsum deposits show that we have to deal with true chemical deposits, and that sometimes evaporation occurred very likely in lagoons shut off from the great lake. Near Geneva, at the Nant d'Avanchet, the following sequence of beds may be observed, from top to bottom :

3. Molasse with gypsum.

2. Fresh-water limestones.

1. Red and violet Molasse and marls.

It is the history of the transformation of a lake owing to evaporation. The red and violet colours of the lower beds show that iron oxides have been washed away from a disintegrated land surface. The deposition of limestones proves a concentration of the water of the lake, and we arrive at the second phase of concentration, the sulphate phase with deposits of gypsum. The Molasse associated with the gypsum beds is made up of sand blown by winds. The last phase of chemical deposits, the chloride phase, is not represented, but chlorides may have been deposited and later removed in solution by water.

3. **Upper marine Molasse.** Marine beds with *Pecten præscabriusculus* overlie the lower fresh-water Molasse. The

upper part represented by the Molasse of Bern and St. Gall with *Ostrea crassissima* has been taken as a type of the *Helvetian*.

4. **Upper fresh-water Molasse.** The fauna of vertebrates found in the upper fresh-water Molasse shows that we have to deal with beds belonging to the Vindobonian, but to the upper part of it, the *Tortonian*.

THE CONGLOMERATES OF THE MOLASSE OR NAGELFLUH

Two kinds of Nagelfluh may be considered :
1. The Juranagelfluh.
2. The Alpine Nagelfluh.

1. **The Juranagelfluh** has been deposited in the Molasse basin along the Jura Range. Its pebbles belong to limestones of the Jura and to the Triassic and crystalline rocks of the Black Forest, showing that they have been transported by streams coming from the north.

2. **The Alpine Nagelfluh** has been deposited on the Alpine border of the Molasse basin. Let us study this category, for its pebbles will allow us to get an idea of the power of erosion at the surface of the Alps recently formed and still moving.

The Alpine Nagelfluh must be divided into :
1. The " Kalknagelfluh " or calcareous Nagelfluh.
2. The " bunte Nagelfluh " or polygenous Nagelfluh.

Generally the " Kalknagelfluh " occurs in the lower beds and the " bunte Nagelfluh " at the top.

If we summarize, after Albert Heim, the results arrived at, we see that *we do not find* in the bunte Nagelfluh pebbles belonging : (1) to the autochthonous crystalline massifs (Mont-Blanc, Aiguilles Rouges, Aar, Gastern, Erstfeld, Gothard) and their sedimentary cover ; (2) to the High Calcareous Alps (Helvetian nappes).

The pebbles of the bunte Nagelfluh are derived from the higher tectonic elements of the Alps, viz., the nappes of the Prealps and of the Austrides. Outcrops where the rocks represented in the pebbles may be found are actually situated 50–150 kilometres to the south and the east.

TECTONICS

The main feature of the Molasse, from a tectonic point of view, is the existence of an anticline, called by Albert Heim the " principal anticline," extending from the Lake of Constance to the Lake of Geneva, over a distance of about 250 kilometres. Two other anticlines are well marked in the eastern part of the Molasse basin, but almost disappear to the west.

The Salève anticline, coming out of the Molasse, near Geneva, made of Jurassic and Cretaceous strata, is very likely the continuation of the " principal anticline " of the Molasse basin. We shall study it soon, for its tectonics and stratigraphy are very interesting.

Let us see now how the marine transgression occurred between the two continental phases, in the history of the Molasse basin.

We have seen above (page 21) that an orogenic paroxysm supervened in the Alps during the middle Oligocene, *but it is not the last movement in the Alps*. Indeed, the folds of the Molasse show that the Miocene beds have been affected by a push coming from the Alps. If we study the region round the Lake of Lucerne and round the Lake of Thun we see that valleys formed by rivers through the Nagelfluh deltas have been invaded in Pliocene times by the last advance of the nappes. We can summarize all these observations in the following moving picture : *the Alps travelling forward, and the pebbles going forward from the chain as it grows and then the chain riding over its own debris.*

The lower Molasse was deposited from the beginning of *Chattian* times in a great lake on the front of the Alps. A slow decrease of the horizontal push made itself felt first at both extremities of the chain, to which corresponds the invasion of the sea in the Rhone basin and in the Vienna basin. In the upper Burdigalian the sinking attained the central segment of the chain. It is the very moment at which the marine basin of Vienna united in front of the Alps with the marine basin of the Rhone. This was the marine transgression. At the Tortonian a renewal of movement occurs in the central segment first, and fresh-water deposits have covered the marine Molasse.

THE SALÈVE (1,308 m.) [1]

Stratigraphy. The oldest strata of the Salève belong to the Kimeridgian Series, the youngest to the Tertiary

[1] Carte géologique du Salève, 1 : 25,000, by E. Joukowsky and J. Favre, Georg. Genève.

(Aquitanian). From a stratigraphical point of view the Jurassic and Cretaceous of the Salève represent the Jura Range type. At the end of the Jurassic we notice an emergence evidenced by the presence of fresh-water beds of the Purbeck.

The following is the sequence of beds exposed :

UPPER MIDDLE OLIGOCENE.—Chattian.—Fresh-water sandstones, grey, more or less marly.

MIDDLE OLIGOCENE.—Marine sandstones.

EOCENE.—Siderolithic. Sandstones with conglomerates at their base.—Continental facies.

CRETACEOUS

Aptian.—Only two outcrops left by the erosion, at Sappey. Rugged yellow limestones, with *Harpagodes pelagi.* (Brongn.), *Ostrea tuberculifera* Koch et Dunk.

Barremian 110 metres.	2.—White reef limestone, Urgonian facies, with *Trigonia ornata* d'Orb., *Arca cornueliana* d'Orb., *Arca dupiniana* d'Orb., *Arca marullensis* d'Orb., *Lima orbignyana* Math., *Terebratula russillensis* de Lor., *Nucleopygus roberti* (A. Gras), *Goniopygus peltatus* Ag., *Cidaris cornifera* Ag.
	1.—Rugged oolitic limestone, yellow, with plenty of sponges mostly indeterminable. *Janira atava* (Roem), *Nucleopygus roberti* (Ag.), *Echinobrissus* aff. *placentula* (Desor), *Pyrina pygaea* (Ag.).
Hauterivian 93 metres.	2.—Zoogenous limestones, microbreccias, oolitic limestones. Equivalent to the Pierre jaune de Neuchatel of the Jura Range. *Toxaster retusus* (Lm.).
	1.—Hauterive marls. *Toxaster retusus* (Lm.), *Exogyra couloni* (Defr.), *Hoplites (Leopoldia) leopoldinus* (d'Orb.), *Hoplites salevensis* Kil., *Hoplites (Acanthodiscus) radiatus* Brug.
Valanginian 33 metres.	2.—Echinodermic limestones, yellowish-brown with *Alectryonia rectangularis* Roem.
	1.—Calcaire roux (yellowish-brown limestone). Quartz and glauconite.
Infravalanginian 98 metres.	2.—Beds with *Natica leviathan* P. & C. Zoogenous microbreccias, marly sandstones with plant remains, oolitic limestones.
	1.—Beds with *Heterodiceras luci* (Defr.). Zoogenous microbreccias, marls with brachiopods, reef-limestones.

JURASSIC

Purbeckian
40–43 metres.

3.—Marls and marly limestones with both marine and fresh-water fossils.
Marine : *Perisphinctes lorioli* Zitt., casts of *Natica* and lamellibranchs at Les Etournelles. At the Petit Salève the fauna is entirely marine or brackish, but all species are dwarfed.
Fresh-water : *Valvata helicoides* Forbes, *Planorbis loryi* Coq., *Physa wealdiensis* Coq., *Lioplax inflata* Sandb., *Chara* sp.
2.—Fresh-water limestones, lithographic or sub-lithographic, sometimes fetid. They contain seeds and fragments of stems of *Chara*, with *Planorbis loryi* Coq., *Physa wealdiensis* Coq., *Lioplax inflata* Sandb., *Cypris* sp.
1.—Marine beds. Oolitic limestones with foraminifera, breccias with pebbles of many colours, surface of desiccation.

Portlandian
65 metres.

2.—Upper. Hard oolitic limestones, forming a cliff. At its base there are spotted limestones.
1.—Lower. Soft reef-limestone forming a ledge. *Diceras speciosum* Munst., *Matheronia salevensis* J. Favre, *Valletia antiqua* J. Favre, *Oonia amygdaloides* (Zitt.).

Kimeridgian
250 metres.

2.—Upper. White reef-limestone.
1.—Lower. Grey limestones, microbreccias with cherts, interstratifications of dolomitic limestones.

Tectonics. From a tectonic point of view the Salève is an elongated anticline occurring in the middle of the Tertiary syncline which separates the Alps from the Jura Range. It is an overfold passing sometimes into a faulted overfold.

The south-eastern limb shelves down gently whilst the north-western limb is vertical. The characteristic feature of the anticline of the Salève is that it has been divided into several compartments owing to tear-faults. There are seven tear-faults giving rise to eight compartments, the eastern members being displaced towards the north-west with reference to the others.

Let us study the tear-fault (*décrochement*) of the Coin which is the most important one. Here the compartment of the Salève has been displaced 700 metres farther to the north-west than the adjacent member of the Pitons, the latter having, in addition, suffered a depression of 80 metres with regard to it.

Though the Petit and Grand Salève are quite distinct from a

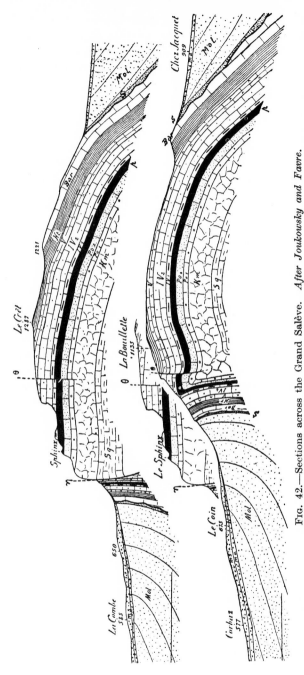

FIG. 42.—Sections across the Grand Salève. *After Joukowsky and Favre.*

Sq, Sequanian. *Km*, Kimeridgian. *Po* 1, Lower Portlandian. *Po* 2, Upper Portlandian. *p*, Purbeckian. *IV* 1, Infravalanginian with *Heterodiceras lucii*. *IV* 2, Infravalanginian with *Natica leviathan*. *V*, Valanginian. *Ht*, Hauterivian. *Bar*, Urgonian. *S*, Siderolithic. *Mol*, Molasse. *ma*, Alpine Moraines. *ms*, Salevian Moraines. *η*, *θ*, Faults.

topographical point of view, they form a geological entity.   Indeed the Monnetier valley is due to erosion and not to a tectonic accident.   The tear-fault of the Coin is well marked in the topography, as shown in Fig. 42.   The deep gullies appearing on the northern cliff are due to transverse faults, of which the vertical throw at the Grande Gorge is 45 metres, while at the Gorge de la Mule the vertical displacement is 35 metres. The gullies of the Sarrot, Varappe and Palavet are fissures filled up with calcite.   We also find longitudinal faults, of which there are four, the best example being seen in the upper part of the Petite Gorge.   Here, owing to this fault, a thin slice, 3 kilometres in length, has been thrown up 50 metres and forms the façade of the mountain.

On the summit of the Salève we find patches of Barremian limestones capped, on the southern slope, by an important outcrop of quartzose sandstones, at times white or stained by red iron oxides. This outcrop has been worked in former times for its iron content, as is shown by small local accumulations of slag. Large erratic boulders of gneiss and granite are present and show that during the Great Ice Age (Riss) the top of the Salève was entirely covered by the glaciers.

From the Salève, if the weather is clear, a magnificent view may be obtained of the northern face of the Alps.   In the foreground the Prealps occur to the left of the river Arve. To the right the High Calcareous Alps, and farther back the Mont-Blanc Massif with its many granite peaks.

The Monnetier valley is due to erosion and represents an old channel of the river Arve. Owing to its capture by another stream, this river now flows in the soft Aquitanian sandstone on the north-eastern extremity of the Salève.

Owing to the pitching towards the north-east, the Barremian, which caps the Grand Salève (1,300 m.), appears at the summit of the Petit Salève (900 m.).

BIBLIOGRAPHY

1. JOUKOWSKY, E. et FAVRE, J.   Monographie géologique et paléontologique du Salève (Haute-Savoie. France).   Mém. Soc. Phys. et Hist. nat. Genève, vol. 37, fasc. 4, Genève, 1913.

2. RÉVIL, J., et MORET, L.   Sur la tectonique de l'axe Salève-Montagne de la Balme-Montagne de Lovagny. Bull. Soc. géol. France, 4e s. t. 22, p. 218, 1923.

3. JOUKOWSKY, E. et FAVRE, J. {Sur les décrochements de la chaîne du Salève. Bull. Soc. géol. France, 4e s. t. XXIV, p. 465–75, 1924.

4. BAUMBERGER, ERNST.—Die Molasse des Schweizerischen Mittellandes und Jura gebietes. Guide Géologique de la Suisse, 1934, p. 57. Wepf et Cie, Bâle.

CHAPTER VIII

THE JURA MOUNTAINS

By Léon W. Collet and A. Werenfels

Introduction and General Outline

The Jura mountains begin in the region of Chambéry, as a virgation of the Alps. Their ranges separate from the Alps and run to the north in two parallel anticlines.

On the northern side of the river Rhone the number of anticlines increases ; they form an arc which is fairly parallel to the Alps and dies away in the Lägern, to the north-east. The span of the arc, taken as a whole, measures about 300 kilometres. The bundle of ranges of the Jura mountains is mostly built up of anticlines, which are like a cloth that has been rumpled up and pushed to one side.

The Jura mountains are undoubtedly the result of a tangential push.

The Central Plateau of France, the small crystalline massif of La Serre, the Vosges and the Black Forest are the obstacles which have prevented the bundle of folds of the Jura from opening to the north like a sheaf. To the east of La Serre and to the south of the Vosges, the folding is not strongly developed, and it is the region in which the folds of the Jura attain their greatest width, 65–70 kilometres. The folding is far more intensified to the south of the Black Forest, in the Hauenstein region, where the folds of the Jura form a bundle pressed together. The direction of the folds is west to east, they hardly measure 10 kilometres in width.

The Vosges and the Black Forest are separated from the folds of the Jura by a broad Table, called the *Jura Tableland*.

The region of the Jura made up of folds is called the *Folded Jura*. As will be seen later on, the boundary between the Jura Tableland and the Folded Jura is marked by an overthrust. In fact, the Folded Jura is thrust over the Jura Tableland.

Where the *Rhine graben* opens a gap of over 50 kilometres in breadth, between the Vosges and the Black Forest, six frontal folds occur. These folds form two arcs with a marked convexity to the north and divide the Jura Tableland into two parts. One table is situated to the west of the trough presented by the Rhine valley, the other to the east of it.

The two parts of the Jura Tableland show the same geological structure, that is a tableland traversed by a system of fractures.

To sum up, the Jura mountains consist of two different parts, from north to south :

The Jura Tableland.

The Folded Jura.

STRATIGRAPHY

Introduction. The sedimentary series of the Jura mountains begins with the Permian, which rests on a crystalline substratum, consisting of the biotite-gneisses of the Black Forest.

The gneisses do not crop out in the Jura mountains, but they are met with, for instance, at Laufenburg, on both sides of the Rhine river, that is on the southern border of the Black Forest. Granite of the Black Forest has been met with in three borings near Schaffhausen, below Rheinfelden and near Zurzach.

If from Laufenburg, we proceed towards the Alps, the first outcrop of gneisses occurs at a distance of 90 kilometres, at Erstfeld in the Reuss valley.

The crystalline series of the Black Forest belongs to an old mountain chain of Hercynian age, which was peneplained.

Carboniferous formations have never been found on the southern border of the Black Forest, though they form a narrow syncline in the Black Forest and crop out at Ronchamps, in the Vosges, where they are exploited.

Permian. The base of the Permian consists of conglomerates and sandstones overlain by dark violet sands, red clays with intercalations of red conglomerates. The upper part is made up of dolomitic beds.

This terrigenous formation very likely represents the *Rotlie-gendes* of the German Permian. Though these strata contain no fossils, they are considered as representing the beginning of a marine transgression.

Trias. In the Jura mountains the *German facies* of the Trias is met with. It is divided as follows, in descending order :

3. Keuper.
2. Muschelkalk.
1. Buntsandstein.

The *Buntsandstein* represents formations deposited on the shore of the continent and on the continental surface in lagoons and great lakes, as shown by Philippi. The red and violet hue of the sandstones indicates a decomposition of the continental surface under a warm and humid climate.

The *Muschelkalk* shows the advance of the sea over the continent In fact, the lower part of the formation is marine. The middle part, on the other hand, indicates a regression of the sea demonstrated by chemical deposits. This middle part of the Muschelkalk is called the *anhydrite group*, which plays a great rôle from an economic point of view as well as from a tectonic point of view. Indeed, it contains rock-salt which is exploited on the southern bank of the Rhine river.

We shall see, later on, that the *anhydrite group* played the rôle of a lubricant during the folding of the Jura, between the overlying and underlying strata, so that the folds of the Jura must be considered, as recorded by Buxtorf, to be due to a phenomenon of " *décollement* ", **that is of ungluing.**

The upper part of the Muschelkalk is made up of marine sediments showing a new transgression of the sea.

The *Keuper* formations indicate a regression of the sea, demonstrated by the presence of sediments deposited in lagoons and lakes, as well as on a sea coast.

In the western part of the Jura the sediments of the Keuper, of continental origin, are overlain by marls and bonebeds which represent a marine transgression. These deposits, only about 1 metre thick, belong to the *Rhaetic stage*, which is considered by the Jura geologists to represent the upper part of the Trias.

Jurassic. During the Jurassic the sea covered the region of the Jura mountains. We have thus to deal with marine deposits. The Ammonites are the fossils which enable us to divide the stages into several zones.

Pioneers such as Leopold von Buch and Quenstedt divided the Jurassic of the Jura as follows :

The *White Jura* or Upper Jurassic
 or Malm . . . . 100–1,000 metres thick.
The *Brown Jura* or Middle Jurassic or Dogger . . . 150– 400 metres thick.
The *Black Jura* or Lower Jurassic
 or Lias . . . . 30– 100 metres thick.

 ———————

 280–1,500 metres thick.[1]

The *Black Jura* consists of clays, marls and shales of a black colour, indicating pelagic conditions.

The *Brown Jura* is made up of sediments deposited in a shallower sea, as demonstrated by the presence of echinodermic limestones and ferruginous oolites. The white colour of the *White Jura* is due to the presence of coral limestones and oolitic limestones deposited under shallow sea conditions.

Generally speaking, we notice a diminution of the depth of the sea, from the Upper Lias up to the Upper Jurassic. At the end of the Jurassic an emergence is indicated by the presence of fresh-water beds of the Purbeck. We noticed the same conditions at the Salève, near Geneva (see page 120).

Lias. In the Jura mountains, the Lias always rests on the Trias. At the Chambelen, to the east of Brugg, on the Reuss river, plant remains and insects have been found in fresh-water formations of the Lower Lias. According to Albert Heim, the land surface from which the plant remains originated was situated to the south. In fact, we know that the Black Forest has been covered by the Liassic sea, while to the south the Aiguilles Rouges and Gastern massifs formed a land surface.

The Liassic formations are generally represented by marls, and shales containing Ammonites which show that Haug's zones are all represented. Detailed stratigraphical studies have not yet been made, and it is not possible, at present, to establish a comparison between the Liassic series of Great

———————

[1] According to Albert Heim.

Britain and the Liassic sequence of the Jura mountains. Lias formations crop out in the eastern part of the Jura mountains, and in the western part of it, on French territory.

Middle Jurassic or *Dogger.* In the western part of the Jura mountains, the *Bajocian* consists of echinodermic and oolitic limestones. In the central part of the ranges ferruginous oolitic limestones are met with, which pass into marls and shales to the east. The *Bathonian* is represented to the west by marly limestones and marls, while in the central region oolitic limestones are met with, as well as farther to the east. The *Callovian* consists of marly limestones to the west, which soon pass into echinodermic limestones.

Upper Jurassic or *Malm.* The *Oxfordian* is represented to the west by marls and to the east by ferruginous oolites. Two different facies of the *Argovian* are met with in the Jura mountains. To the north occur coral limestones which have been called *Rauracian* by the Jura geologists. To the south the facies of the Rauracian passes into marls with Ammonites, which represent the true *Argovian.*

The *Sequanian* is made up of limestones and marls, while the *Kimeridgian* consists of white coral limestones. The *Portlandian* has been removed by erosion in the eastern part of the Folded Jura, while it is represented in the western part by oolitic limestones, microbreccias and dolomitic limestones, a facies very similar to the Portlandian of the Salève (page 120). At Morteau (France) a brackish fauna occurs in the upper beds of the Portlandian, showing the passage into the fresh-water deposits of the *Purbeck.*

Purbeck beds are well exposed in the south-western Folded Jura, as shown by Maillard, Lagotala and Nolthenius. The sequence of the Purbeck beds of the Salève, which belong stratigraphically to the Jura, is given on page 120.

In the Jura Tableland and in the eastern part of the Folded Jura, exposures of all the formations of the Lower and Middle Jurassic are met with. In the central part of the Folded Jura, Upper Jurassic and Cretaceous beds are exposed, while to the south-west in some deep cuttings, due to erosion, Middle Jurassic and even Lias crop out.

To sum up, the Jurassic is represented in the Jura mountains from the Lias up to the Purbeck by marine formations. An emergence is indicated, at the end of the Jurassic, by the presence of fresh-water beds of the Purbeck. The Upper Jurassic was removed by erosion of Cretaceous and Tertiary age in the eastern Folded Jura.

Cretaceous. In the eastern part of the Folded Jura, Cretaceous formations do not occur. They appear to the west of Bienne and develop in thickness to the south-west. A pre-Cenomanian denudation took place in the region of the lake of Bienne. Moreover, another denudation occurred in the Upper Cretaceous and during the Eocene, which explains the absence of the Cretaceous formations in the eastern Folded Jura. Evidence of a third phase of denudation of post-Miocene age has been recorded by Buxtorf.

Neocomian. The classical series of the Lower Cretaceous have been ascertained on the southern margin of the Folded Jura, near Neuchatel (Neocomum). The sediments of the Neocomian were deposited in a shallow sea. They rest on the Purbeck beds. The Neocomian is divided into the two following stages : the *Valanginian* at the base and the *Hauterivian* at the top.

The *Valanginian* has been divided by Baumberger as follows, from top to bottom :

Upper Valanginian.—13 m.

4. " Couche de Villers." Yellow marls with *Exogyra couloni, Toxaster granosus, Saynoceras verrucosum, Astieria atherstoni.*—0·3 m.
3. Thin bedded limestones containing limonite.—3·5 m.
2. " Calcaire roux." Echinodermic limestone.—9·0 m.
1. Yellowish marly limestone, representing the " Marnes d'Arzier " of the south-western Jura. *Rhynchonella valangiensis, Terebratula valdensis, Zeilleria tamarindus.*—0·3 m.

Lower Valanginian or Infravalanginian (Berriasian).— 42–44 m.

2. " Marbre bâtard." Limestones, often oolitic, with intercalations of marls. *Natica leviathan, Natica pidanceti, Terebratula valdensis, Nerinea valdensis.*—20 m.
1. Grey marls and oolitic limestones. *Pygurus gillieroni, Phyllobrissus duboisi, Terebratula valdensis.*

The *Hauterivian* consists of four parts, which are from top to bottom :

4. " Pierre jaune de Neuchatel." Yellow oolitic limestone with intercalations of echinodermic limestones.—22 m.

S.A.

3. Yellow bedded oolitic and echinodermic limestone with glauconite and siliceous limestones concretions. *Toxaster complanatus, Holectypus macropygus, Rhynchonella multiformis.*—12 m.

Marnes d'Hauterive
$\left\{ \begin{array}{l} \text{2. Grey marls with calcareous concretions.} \\ \quad \textit{Hoplites (Leopoldia) leopoldinus, Nautilus} \\ \quad \textit{neocomiensis, Exogyra couloni.}\text{—3 m.} \\ \text{1. Blue marls.—4 m.} \end{array} \right.$

Ammonites have also been found in the " Pierre jaune de Neuchatel," but they are not so common as in the marls.

The *Barremian* is represented by yellow oolitic limestones at the base and by white coral limestones at the top. This facies is known as *Urgonian*. In the " Val de Travers," the Urgonian contains asphalt which is exploited. It is possible that the upper part of the Urgonian represents the lower part of the *Aptian*. In this stage, marls with *Orbitolina*, and sandstones with glauconite are met with.

The *Albian* is mostly made up of fossiliferous green sands (glauconitic) often containing phosphatic concretions. The classical exposure of the Albian of the Jura is at Bellegarde, on the banks of the Rhone, where detailed studies have lately been made by Jayet. Another outcrop of Albian occurs at Ste. Croix. Farther to the east, the Albian has been removed by erosion.

The *Cenomanian* consists of chalky limestones and marls in the region of the lake of Bienne, where it rests on Lower Cretaceous or even on the Upper Jurassic. This fact demonstrates a pre-Cenomanian denudation, referred to above. At Bellegarde green sands without fossils represent the Cenomanian.

The *Turonian* is only met with in small outcrops in synclines of the south-western Jura. It consists of sublithographic limestones with *Rosalina linnei*. Pebbles of these limestones occur in the Eocene, showing that the Turonian has been disintegrated during a continental phase.

Tertiary. Outcrops of Tertiary formations occur in the synclines of the Folded Jura and in the Jura Tableland. We have to deal with two different facies : continental and marine.

Eocene. Between the sediments of the *Molasse* and the Mesozoic sediments of the Jura mountains occurs a continental formation made up of ferruginous clays and quartz sands called " siderolithic " or iron-stone. Vertebrate remains show that this formation has been formed under a tropical climate, during the Middle and Upper Eocene.

Oligocene. In the Delémont basin occur, on the Siderolithic deposits of gypsum, fresh-water limestones and conglomerates of *Sannoisian* age (Lower Oligocene). With the Middle Oligocene, *Stampian*, appears the *Molasse*, in two different facies : a marine facies to the south and west of the upper Rhine plain and a fresh-water facies occurring in the Folded Jura. The Upper Oligocene or *Aquitanian* consists also of fresh-water *Molasse*, called the " Molasse rouge " owing to its red colour due to iron oxide pigmentation (see page 116).

Miocene. The Miocene formations of the Jura belong to the marine Molasse and to the Upper fresh-water Molasse (see page 116). The marine Molasse begins with the *Burdigalian* and is transgressive on the Red fresh-water Molasse. The Middle Miocene or *Helvetian* is made up of polygenous conglomerates deposited on the sea margin. In two localities of the eastern Jura shore deposits occur, as at the Tennikerfluh and at the Randen. During the Upper Miocene or *Tortonian* a regression of the sea took place to the west. In the eastern and central Folded Jura fresh-water formations occur, while in the western part marine sediments were deposited at the same time. In the eastern Jura Tableland fluviatile conglomerates occur which pass into the lacustrine facies of the " Oehningerkalk." In the Folded Jura fluviatile sands containing remains of *Dinotherium giganteum* pass also into the fresh-water beds of Oehningen.

Pliocene. The youngest formations of the Tertiary belong to the Lower Pliocene or *Pontian*. They consist of fluviatile pebbles and sands brought down from the Vosges and the Black Forest by streams. These Pontian formations fill up the western part of the Delémont basin and cap the highest ridges to the south of the Rhine graben.

TECTONICS

I. THE JURA TABLELAND

Great or Eastern Jura Tableland (Basel and Aargau). The Great or Eastern Jura Tableland represents a table made up of Triassic and Jurassic formations dipping at an angle of 1°–10° towards the south and south-east. This table is divided into several blocks by faults and flexures. Folds may also occur, as well as valleys due to erosion, which may form small tables.

Buxtorf recorded, for the first time, that the planes of the boundary-faults of a graben may converge downwards, so as to form a **wedge-graben** (Fig. 48). The faults of the Jura Tableland are not only marked in the Trias but also in the crystalline substratum, so that they must be due to movements of the crystalline massif (Fig. 44).

As pointed out by Buxtorf, the Siderolithic and Oligocene formations were affected by the faults, while the Helvetian and Tortonian Molasse were not disturbed. As the same fact occurs in the case of the boundary-faults of the Rhine graben, we may conclude that the formation of the Rhine graben and of the faults of the Jura Tableland are of the same age.

In Buxtorf's opinion, the *wedge-grabens* are due to the obstacle presented by the Black Forest to the development of the orogenic push coming from the Alps.

Fig. 48, after Buxtorf, shows the formation of the Great Jura Tableland from Eocene till recent times.

The Western Jura Tableland. The Western Jura Tableland is situated to the south of the Vosges and to the west of the *Rhine graben*, in the triangle marked by the towns of Belfort-Besançon-Porrentruy. The surface of the Table is carved-out Jurassic formations, from the Oxfordian up to the Upper Jurassic.

II. The Folded Jura.

The "Décollement" of the Folded Jura. As shown by Buxtorf, the folding of the Jura must be considered as a *surface phenomenon.* In fact, crystalline rocks of the substratum, Permian and even the oldest strata of the Trias (Buntsandstein) do not occur in the core of the anticlines, *for they have not been folded.* The oldest folded strata met with belong to the Middle Muschelkalk, that is to the Anhydrite formation, which contains beds of rock-salt.

Owing to the push exerted by the Alps in formation, the sedimentary rocks of the Jura have been detached on the Middle Muschelkalk, of which the salt beds played the rôle of a lubricant, and folded. This kind of folding is a "*décollement*" (Fig. 43).

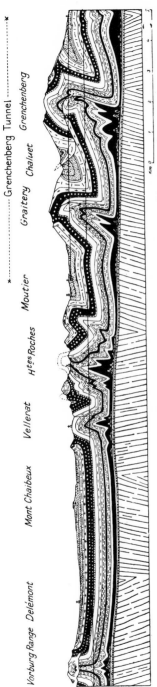

Grenchenberg Tunnel

FIG. 43.—Section showing the " décollement " of the Folded Jura. *After Buxtorf.*

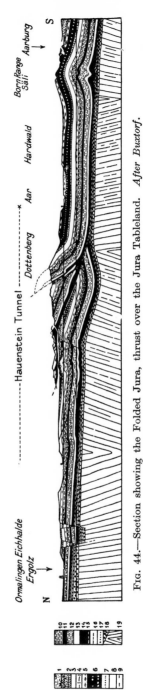

Ormalingen Eichhalde
Ergolz
N

Hauenstein Tunnel

Dottenberg Aar   Hardwald   Born Range
                                   Sälì   Aarburg
                                             S

FIG. 44.—Section showing the Folded Jura, thrust over the Jura Tableland. *After Buxtorf.*

1, Quaternary. 2, Molasse. 3, Portlandian. 4, Kimeridgian. 5, Sequanian. 6, Argovian. 7, Upper Dogger. 8, Middle Dogger. 9, Lower Dogger.
10, Opalinus clays. 11, Lias. 12, Keuper. 13, Upper Muschelkalk. 14, Anhydrite formation. 15, Beds of rock-salt. 16, " Wellenkalk." 17, Buntsandstein.
18, Permian. 19, Crystallines of Hercynian age (Black Forest).

133

That we have to deal with a surface phenomenon is demonstrated by many anticlines which pitch out, and are replaced by another anticline emerging from the bordering syncline (Fig. 45).

The different kind of folds. In general in the Jura mountains, every range represents an anticline. Symmetrical anticlines represent the simplest folding. When the push,

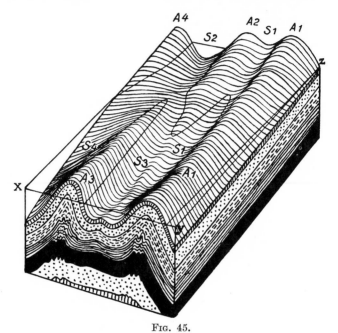

FIG. 45.

A1–A4, Anticlines. S1–S4, Synclines. A3, Anticline pitching out and replaced by A2.
After B. G. Escher (*Algemeene Geologie*).

to which the folding is due, increases, symmetrical anticlines pass into more complicated folds.

The first complication occurs in the archbend of the anticline, which may be divided into two secondary anticlines, as shown in Fig. 46.

In some cases a **trunk-anticline** may originate, in which both limbs are very steep or even vertical. A good example is shown in the Clos du Doubs range and in the Cornu-Foulet anticline (Fig. 47). If the push continues, one of the limbs may form an overfold. When both limbs form overfolds in opposite directions, we have to deal with a *fan-shaped fold*. In some cases, by the exaggeration of the push at the base of the fold, a *detached arch core* may be met with.

A fault may complicate the overfolded limb of a fan-shaped fold, as shown in the Saignotte anticline (Fig. 47).

To the trunk-anticline corresponds a trunk-syncline, which is called a *basin* (cuvette). The trunk-anticlines are seldom met with in the eastern ranges of the Jura, while they occur in the south-western part.

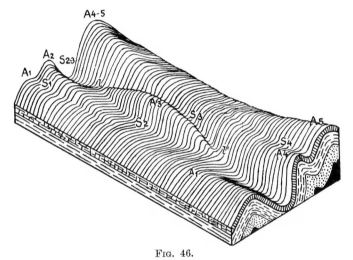

FIG. 46.

A1–A5, Anticlines. *S1–S4*, Synclines. Complications in the archbend of an anticline. *After B. G. Escher (Algemeene Geologie).*

A *symmetrical anticline* may pass into an *unsymmetrical anticline*, in which one limb is steeper than the other. By increase of the push, this fold passes into an overfold. By exaggeration of the movement a *faulted overfold* (Fig. 47) is met with.

Clean cut thrusts occur also in the Jura mountains. This type of thrusts appears at the front of the Folded Jura, where the external anticline is thrust over the Jura Tableland. In this case the Anhydrite formation of the Trias rests on the Tertiary of the Table. This region represents the principal zone of overthrusts of the Jura mountains. To the east, this zone is not so well developed and dies away in the Lägern range.

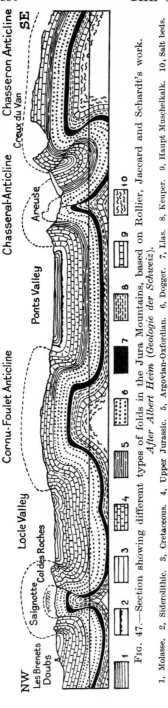

FIG. 47.—Section showing different types of folds in the Jura Mountains, based on Rollier, Jaccard and Schardt's work. After Albert Heim (*Geologie der Schweiz*).

1, Molasse. 2, Siderolithic. 3, Cretaceous. 4, Upper Jurassic. 5, Argovian-Oxfordian. 6, Dogger. 7, Lias. 8, Keuper. 9, Haupt Muschelkalk. 10, Salt beds.

Secondary thrusts may develop parallel to the principal thrusts, so that a *schuppen structure* may occur (Fig. 44) as in the case of the Dottenberg.

Owing to erosion, parts of a slice of the Folded Jura, thrust over the Jura Tableland, may be isolated so as to form an outlier or a *klippe*. They are not rare on the boundary between the Jura Tableland and the Folded Jura. Some of the klippes occur on the Jura Tableland, at a distance of 5 kilometres from the Folded Jura. The type of the overthrusts which occur on the boundary between the Folded Jura and the Jura Tableland as well as in internal parts of the ranges, belongs to the *clean cut thrusts* (Fig. 44).

The boring of the railway tunnel Moutier-Granges, which over a distance of 8,565 metres traverses two of the internal anticlines of the Folded Jura, enabled Buxtorf to make interesting observations. In fact, this author found that *folded clean cut thrusts* occur in the internal part of the Jura range.

In Buxtorf's opinion a fracture, the plane of which was not steep, was the beginning of the phenomenon. Owing to the Alpine push, the upthrown block advanced towards the north and developed into a thrust mass. Then the latter and its substratum were folded (Fig. 43).

Direction of push of the Overfolds. As shown by Buxtorf, four-fifths of the Folded Jura consist of overfolds trending to the north, that is to its external part. In some cases, especially on the southern border of the Folded Jura, overfolds directed to the south are met with. They represent a phenomenon of backward folding, due to the fact that the northern part of the Molasse was pushed underneath the southern foot of the first range of the Folded Jura, the upper part of which had to fold back. In Buxtorf's opinion, the backward folding represents one of the latest phases of the formation of this mountain, as it occurred in the case of the Alps for the backward folding or fan-shaped structure of the roots. The backward folding of the Folded Jura may thus correspond to Argand's *phase insubrienne* of the Alps formation.

Disharmonic Folding. Soft strata, such as shales and marls intercalated between hard formations, may play the rôle of a lubricant during the folding, producing disharmonic folding (see page 12). As shown by Buxtorf, disharmonic folding often occurs in the Folded Jura. In fact, the lubricant is represented by the Upper Trias (Keuper), the lowest part of the Middle Jurassic and the Oxfordian-Argovian. The latter formation especially produced disharmonic folding. A very good example is shown in the syncline (Fig. 43) situated to the south of the Vellerat anticline. The Upper Jurassic forms a syncline on the Oxfordian clay, while the Middle Jurassic (Dogger) forms an anticline underneath the Oxfordian clay. Such structures are not important and must be considered as accidents in the folded system of the Jura mountains.

Tear-faults. In the western part of the Folded Jura there is no evidence of fractures formed before the main folding, but 8 tear-faults, giving rise to 9 compartments, occur between the Salève and Neuchatel. They diverge towards the north.

The eastern compartments are generally displaced towards the north with reference to the others. These tear-faults are due to the bow structure of the Folded Jura and thus produced an elongation of it. The horizontal displacement may vary from 300 metres to 10 kilometres. The total elongation due to the tear-faults measures 10 kilometres. The length of the Jura bow which, according to Albert Heim, was formerly of 320 kilometres, thus attains a length of 330 kilometres.

The greatest tear-fault of the Folded Jura occurs between Vallorbe and Pontarlier, over a distance of 45 kilometres. It is well marked in the scenery as a great fracture cutting the folds at an angle of 55°.

The tear-fault of St. Cergue, studied and mapped by Lagotala, shows a horizontal displacement of the western compartment with reference to the eastern. The elongation of the Jura range at St. Cergue, owing to the tear-fault, measures 600–650 metres.

Disposition of the Folds. The Folded Jura is characterized by the fact that the highest folds are situated at its internal border and the lowest at its external margin. The folds developed better in ranges to the south-west and north-east, where they form bundles. In the middle part of the Folded Jura, the folds are wider and lower. There are about 160 anticlines, from a few kilometres in length to about 30 kilometres. The overthrust zone of the frontal part of the Folded Jura may be followed over 160 kilometres.

The Folded Jura may be divided into two zones :

(*a*) The Internal Zone.

(*b*) The External Zone.

The *Internal Zone*, about 20 kilometres in width, borders the Swiss Plateau (Molasse). It consists of the eastern Folded Jura, characterized by overthrusts and "*schuppen structure*," then of the Berner Jura made up of symmetrical anticlines, while the south-western part of the zone contains trunk-folds and fan-shaped folds.

The *External Zone* shows lower folds, more distant from one another, with anticlines separated by large synclines.

DIRECTION AND AMOUNT OF THE TANGENTIAL
COMPRESSION

The *direction of the orogenic forces* is inferred from the convexity of the arcs formed by the folds. In the Jura mountains it is easy to determine that the convexity of the arcs is directed towards the north, showing that the push came from the Alps, through the Tertiary syncline of the *Molasse*.

The amount of the compression due to the folding of the Jura mountains is greatest to the east of the Rhine graben and measures 17·5 kilometres. For a single anticline the compression may vary from 250 metres to 7 kilometres.

Scenery of the Folded Jura

The scenery of the Folded Jura harmonizes with its geological structure. The anticlines are ridges, while the synclines are longitudinal valleys. Closer study, however, shows that the Jura mountains have lost, at least, half of their primitive height, owing to erosion.

The **synclinal valleys** of the Folded Jura are quite conspicuous. In the Western Jura the trough core is generally *Molasse* and Cretaceous. Main streams follow synclinal valleys for short distances only.

Synclinal ridges, due to an inversion of the relief, are seldom met with in the Jura mountains.

Isoclinal valleys, which are the result of denudation, may be well developed. Indeed, some of them are large enough to contain hamlets or villages ; generally they are but narrow vales without any water-courses. Such valleys are named *combes.* They owe their origin to soft formations intercalated between hard strata.

Anticlinal valleys are often met with. They were carved out of anticlinal ridges, and represent phenomena due to the erosion, not to the rupture, of the arch of an anticline. Anticlinal valleys are not of great length and are always connected with transverse valleys, which were formed first. Thus they represent small affluent valleys of the latter.

That the anticlinal valleys are not due to longitudinal fractures is shown in the upper part of their catchment basin, where the complete archbend of the anticline is obvious. The archbend of the Jura anticlines never consists of Cretaceous strata, for they have been eroded away. The younger rocks of the archbends belong to the Upper Jurassic, the older to the Trias, as is shown in the eastern part of the Jura, as well as in its western part (France).

Anticlinal ridges and isoclinal ridges characterize the scenery of the Folded Jura. The valleys developed among these folds show a juvenile nature.

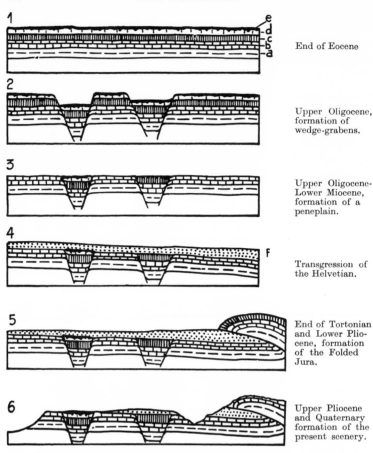

Fig. 48.—The geological History of the Jura Tableland, near Basel.
After A. Buxtorf.

a, Lower Dogger. *b*, Middle and Upper Dogger. *c*, Lower Malm. *d*, Middle Malm.
e, Eocene (ironstone). *f*, Helvetian-Tortonian (Molasse). Dogger = Middle Jurassic.
Malm = Upper Jurassic.

The **transverse valleys** or **cluses** are also an important
feature of the Folded Jura. They are deep cuttings across
anticlines, in which the internal structure of the fold is exposed.
They are due to erosion. To explain the formation of such
cluses, we must consider that the river is older than the anticline and that the erosion took place *pari passu* with the
uplift of the anticline. According to Albert Heim, there are
60 to 70 *cluses* in the Folded Jura. One of the most interesting
occurs between Moutier and Delémont. It is so narrow that

there is only room for the Birs river, the road and the railway line which in some places passes in tunnels.

Gullies and transverse valleys, which are younger than the anticlines, are also met with. A stream developed on the limb of an anticline and owing to head erosion worked its way farther and farther back into the anticline until the latter is crossed. Such cases are rare in the Folded Jura.

Transverse valleys due to Tear-Faults. These valleys are characterized by the fact that they are oblique to the direction of the folds. Tectonic features never correspond on both sides of the valley, owing to the tear-fault. The river is independent of the valley.

These valleys are naturally of advantage when building roads and railway lines. In the regions where they do not occur, as in the eastern part of the Folded Jura, tunnelling is necessary.

BIBLIOGRAPHY

1. BUXTORF, A.—Geologische Beschreibung des Weissensteintunnels und seiner Umgebung. Beitr. z. geol. Karte d. Schweiz N.F. XXI. Bern, 1908.
2. BOURGEAT.—Sur les failles courbes des environs de Salin (Jura). Bull. Soc. géol. France. 1908.
3. WERVECKE, L. VAN.—Die Tektonik des Sundgaues und ihre Beziehungen zur Tektonik der angrenzenden Teile des Juragebirges. Mitteil. d. geol. Landesanst. Elsass-Lothringen, 6, 1909.
4. VERLOOP, J. H.—Die Salzlager der Nordschweiz, Diss. Basel, 1909.
5. BLOESCH, E.—Zur Tektonik des schweizerischen Tafeljura. Neues Jahrb. Min. Beil.-Bd. XXIX. 1910.
6. BUXTORF, A.—(a) Oberflächengestaltung und geol. Geschichte des nordschweizerischen Tafeljura. (b) Analogien im Gebirgsbau des schweiz. Tafeljura und der arabischen Wüste. Mém. Soc. helv. Sc. Nat. I. 1910.
7. CLOOS, H.—Tafel- und Kettenland im Basler Jura und ihre tektonischen Beziehungen, etc. Neues Jahrb. Min. etc. Beil.-Bd. XXX. 1910.
8. BRÄNDLIN, E.—Zur Geologie d. nördlichen Aargauer Tafeljuras zwischen Aare und Fricktal. Basler Verh. XXII. 1911.
9. KEMMERLING, G.—Geologische Beschreibung der Ketten von Vellerat & Moutier. Inaug.-Diss. Freiburg i/Br. 1911.
10. DELHAES, W., und GERTH, H.—Geol. Beschreibung des Kettenjura zwischen Reigoldswil (Baselland) und Oensingen (Solothurn). Geol. & Pal. Abh. herausg. v. E. Koken, N.F. XI. Heft I. 1912.

11. HUMMEL, K.—Die Tektonik des Elsgaues. Berichte d. Nat. Ges. zur Freiburg i/Br. Bd. 20. 1914.
12. WILSER, J. L.—Die Rheintalflexur nordoestl. von Basel zwischen Lörrach und Kandern und ihr Hinterland. Mitteilung der badischen Geologischen Landesanstalt, Heft 2. 1914.
13. DISLER, C.—Stratigraphie u. Tektonik des Rotliegenden und der Trias beiderseits des Rheines zwischen Rheinfelden und Augst. Basler Verh. xxv. 1914.
14. AMSLER, A.—Tektonik des Staffelegg-Gebietes. Eclogae geol. Helvet. xiii. 1915.
15. HEIM, ALB.—Die horizontalen Transversalverschiebungen im Juragebirge. Vierteljahrschrift der naturf. Ges. Zürich, 1915.
16. SUTER, R.—Geologie der Umgebung von Maisprach (Schweizerischer Tafeljura). Basler Verh. xxvi. 1915.
17. BUXTORF, A.—Prognosen und Befunde beim Hauensteinbasis &. Grenchenbergtunnel und die Bedeutung des letztern für die Geologie des Juragebirges. Basler Verh. xxvii. 1916.
18. HEIM, ALB.—Geologie der Schweiz. Band i. Leipzig, 1919. C. H. Tauchnitz.
19. STRIGEL, A.—Ueber prätriadische Einebnung im Schwarzwald. Jahresber. u. Mitt. d. Oberrh, geol. Ver. N.F. viii. 1919.
20. LAGOTOLA, H.—Monographie géologique de la région de la Dôle— St. Cergue. Mat. Carte géol. Suisse. N.S. 16. Berne, 1920.
21. ELBER, R.—Geologie der Raimeux- und der Vellerat-kette im Gebiet der Durchbruchstäler von Birs & Gabiare (Berner Jura). Basler Verh. xxxii. 1920.
22. GRAHMANN, R.—Der Jura der Pfirt im Oberelsass. Ein Beitrag zur Kenntnis der Geschichte des Oberrheintalgrabens. Neues Jahrb. Min. Beil.-Bd. xliv. 1920.
23. LEHNER, E.—Geologie der Umgebung von Bretzwil im nordschweizerischen Juragebirge. Beitr. z. geol. Karte d. Schweiz. N.F. 47. ii. Bern, 1920.
24. SUTER, HANS.—Geolog. Untersuchungen in der Umgebung von Les Convers-Vue des Alpes. Diss. Zürich, 1920.
25. AMSLER, ALFR.—Beziehungen zw. Tektonik und tertiärer Hydrographie im oestl. Tafeljura. Eclogae geol. Helvet. xvi. 1922.
26. KELLER, W. T.—Geologische Beschreibung des Kettenjura zwischen Delsbergerbecken und oberrheinischer Tiefebene, etc. Eclogae geol. Helvet. xvii. 1922.
27. MARGERIE, EMMANUEL DE.—Le Jura. Mémoires pour servir à l'explication de la carte géologique détaillée de la France. Paris, 1922.
28. KOCH, R.—Geologische Beschreibung des Beckens von Laufen im Bernerjura. Beitr. z. geol. Karte d. Schweiz. N.F. 48. Bern, 1923.
29. SENFTLEBEN, GERH.—Beiträge zur geologischen Erkenntnis der West-Lägern und ihrer Umgebung. Diss. Zürich, 1923.
30. WIEDENMAYER, C.—Geologie der Juraketten zwischen Balsthal und Wangen a/Aare. Beitr. z. geol. Karte d. Schweiz. N.F. xlviii, 3. Bern, 1923.
31. STAEHELIN, P.—Geologie der Juraketten bei Welschenrohr, Kanton Solothurn. (Abschnitte der Raimeux-Kette, etc.) Beitr. z. geol. Karte d. Schweiz. N.F. lv. Liefg. 1. Bern, 1924.
32. BIRKHÄUSER, M.—Geologie des Kettenjura der Umgebung von Undervelier (Berner Jura). Basler Verh. xxxvi. 1924/25.

33. FREI, E.—Zur Geologie des südoestl. Neuenburger Jura. Beitr. z. geol. Karte d. Schweiz. N.F. LV, 3. Bern, 1925.
34. WAIBEL, A.—Geologische Beschreibung des Kartengebietes von Blatt Erschwil. Beitr. z. geol. Karte d. Schweiz. N.F. LV, 2. Bern, 1925.
35. VOSSELEŘ, P.—Die tertiäre Entwicklung des Aargauer Tafeljura. Mitt. d. Aarg. Natf. Ges. Heft 17. 1925.
36. AMSLER, A.—Bemerkungen zur Juratektonik (Tektonische Jura-probleme). Eclogae geol. Helvet. XX. 1926.
37. LINIGER, H.—Ueber Gitterfaltung im Berner Jura. Eclogae geol. Helvet. XIX. 1926.
38. LINIGER, H., und ⌠Zur Tektonik der Umgebung von Asuel und WERENFELS, A. ⌡St. Ursanne. Eclogae geol. Helvet. XX. 1927.
39. FAVRE, JULES, et ⌠Le Jura. Guide géologique de la Suisse, JEANNET, ALPHONSE.⌡1934, p. 42. Wepf et Cie, Bâle.
40. BAUMBERGER, ERNST.—Die Molasse des Schweizerischen Mittel-landes und Jura gebietes. Guide géologique de la Suisse, 1934, p. 57. Wepf et Cie, Bâle.

PART III

THE GEOSYNCLINE, THE WESTERN ALPS OR THE PENNINE NAPPES

CHAPTER I

INTRODUCTION

We have seen (page 22) that the Pennine Nappes were formed in the Alpine geosyncline. Argand has shown that we have to deal with six nappes which are numbered as follows, from top to bottom :

VI. The Dent Blanche Nappe.
V. The Monte Rosa Nappe.
IV. The Great St. Bernard Nappe.
III. The Monte Leone Nappe.⎫
II. The Lebendun Nappe. ⎬Simplon-Ticino Nappes.
I. The Antigorio Nappe. ⎭

When studying the structural divisions of the country, we saw that the Simplon region represents a zone of culmination, which means that from this region there is an outward pitching, towards the south-west, and towards the north-east.

This very important culmination extends from the Aosta Valley, on the south-west, to the valley of Poschiavo, on the north-east—a distance of about 220 kilometres. Argand discovered the nappes of the Pennine Alps to the south-west of the culmination, and on the basis of this discovery Staub found their homologues on the north-east of the culmination.

The correlation of the Pennine Nappes, on both sides of the culmination, is as follows :

South-West.	*North-East.*
VI. Dent Blanche Nappe .	Margna Nappe.
V. Monte Rosa Nappe . .	Tambo-Suretta Nappe.
IV. Great St. Bernard Nappe .	Adula Nappe.
I–III. Simplon-Ticino Nappes	Simplon-Ticino Nappes.

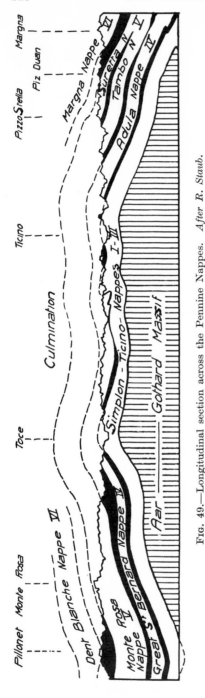

FIG. 49.—Longitudinal section across the Pennine Nappes. *After R. Staub.*

Fig. 49 shows the disposition of the nappes on both sides of the culmination, in a longitudinal section.

Thus in the Pennine Alps, stretching from the Simplon to the Dora Baltea, we have an axis depression from northeast to south-west. The deepest nappes crop out in the Simplon-Ticino region ; while in the southwest part of the Pennine Alps, viz. in the area of the Dora Baltea, the highest nappes occupy the surface, and the deeper nappes are not to be seen at all.

In order to grasp more thoroughly the importance of the Simplon-Ticino culmination, let us start from Gondo, on the southern slope of the Simplon, to reach Arolla. We shall pass over the Portjengrat, through Almagell and Mattmark, on the Schwarzenberg-Weissthor, through Zermatt and over the Col d'Hérens.

Our trip leads us onwards, as if it were a giant staircase, from the lowest nappe to the highest, and we climb the 25 kilometres in which they are piled up. At our starting-point, above us, 25 kilo-metres of rocks have been carried away by erosion; but less and less material is missing, as we proceed towards the south-west. Does this mean that the Alps once reached such heights ? Probably not, for erosion and isostatic adjustments were at work to lower them, concurrently with their successive uprisings.

CHAPTER II

THE SCHISTES LUSTRES

Introduction

From a petrological point of view, the *Schistes lustrés* represent the deep-sea sediments deposited in the Alpine geosyncline, or Tethys.

From a tectonic point of view, they fill up the synclines of the Pennine Nappes and also surround their frontal archbends where these have not been removed by erosion. They constitute the envelope of the Paleozoic cores of the Pennine nappes and separate these cores from the Foreland. They may be followed from the Mediterranean Sea as far as the Rhone valley, and outcrop behind the Hercynian massifs of the Mercantour, Pelvoux, Belledonne, and Mont-Blanc.

They are exposed along the Rhone valley, behind the High Calcareous Alps and the Aar Massif. They then continue at the back of the Gothard Massif as far as the river Rhine, and the foot of the Silvretta.

The *Schistes lustrés* are nothing but the " *Bündner Schiefer* " of the German Swiss geologists and the " *Zona delle pietre verdi* " of the Italian School.

Some of the massifs of the Pennine Alps, such as the Gran Paradiso and Ambin, seem to emerge from under the *Schistes lustrés*, so that, before Argand's discoveries in the region of Monte Rosa, they were taken for massifs rooting *in situ*.

If we consider the highest of the Pennine tectonic elements, the Dent Blanche Nappe, we see it, separated from its roots by erosion, **floating on the Schistes lustrés.** This fact, of paramount importance, is to be observed from the Gornergrat and from the Dora Baltea valley. Indeed, this phenomenon is so great that many geologists who accompanied me to the Gornergrat had difficulty in realizing that the nappes of the

148

Pennine Alps were on such a scale. When we see the Matterhorn, Dent Blanche, Ober Gabelhorn, Rothorn, Weisshorn, all made up of Paleozoic crystalline rocks resting on a ledge of *Schistes lustrés*, we understand the full value of Argand's work (Fig. 50).

Those who have followed the path leading from les Chapieux, in France, to Courmayeur, in Italy, through the Col de la Seigne, on the southern side of Mont-Blanc, have realized the dull and monotonous scenery of the *Schistes lustrés* that form the range in which Mont Favre and Crammont are the principal summits. It is a contrast to find, farther to the east, a huge mountain like the Grand Combin carved out of these same *Schistes lustrés* where they make the substratum of the Dent Blanche Nappe.

Where the Pennine Nappes have been thrust against the Hercynian massifs of the Foreland, the *Schistes lustrés* are piled up against the obstacle, as in the region between Les Haudères and Arolla. Where, on the other hand, the Pennine Nappes have travelled without hindrance towards the north, the *Schistes lustrés* are very thin, as is conspicuous in the substratum of the Weisshorn and of the Diablons.

PETROLOGY OF THE SCHISTES LUSTRÉS

When dealing with the petrology of the *Schistes lustrés* we must consider that the depth of the

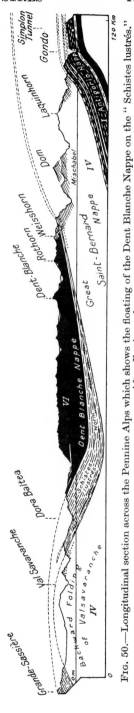

FIG. 50.—Longitudinal section across the Pennine Alps which shows the floating of the Dent Blanche Nappe on the "Schistes lustrés." *After E. Argand.*

geosyncline in which they were deposited, was not every-
where the same. Indeed, near the Foreland and on both
sides of the geanticlines and cordilleras, shallow-water deposits
must be expected (see page 20), as well as chemical deposits
like gypsum, while in the deep part of the secondary geosynclines deep-sea deposits occur.

The *Schistes lustrés* are more metamorphosed in the region
of the deepest Pennine Nappes of the Simplon-Ticino group,
than in the higher nappes.

Intercalations and sills of greenstones are common and they
increase in quantity towards the south.

The *Schistes lustrés* may be divided into two groups, as
follows :

1. The Gothard group.
2. The Pennine group.

1. **The Gothard group** is made up of sediments deposited
on the southern side of the Hercynian Massif of the Gothard.
In spite of the metamorphism, it is easy to recognize deposits
containing more terrigenous constituents than occur in the
Schistes lustrés of the Pennine group.

To the Gothard group belong the *Schistes lustrés* of the Piora
syncline, situated between the Lucomagno Massif and the
Gothard Massif. The *Schistes lustrés* of the Nufenen zone
belong to the same group.   The *Schistes lustrés* are here characterized by the presence of the upper Trias at their base,
represented by white bedded quartzites, passing into micaschists
and black phyllites with garnets.   These schists were formerly
like the marls and clays, with iron oxides, which characterize
the upper Trias associated with the Hercynian massifs.   Black
calcareous phyllites, containing intercalations of sandstones,
overlie these formations.   Their black colour is due to a
carbonaceous pigment.

2. **The Pennine group.**   These *Schistes lustrés*, with the
exception of those deposited near the geanticlines, represent
a deeper facies.   They are generally distinguished from those
of the Gothard group by their grey hue, for they do not contain
much carbonaceous pigment.

According to Preiswerk, we may divide the *Schistes lustrés*
of the Simplon-Ticino Nappes into the following categories :

(*a*) Calcareous phyllites.
(*b*) Grey and black phyllites with garnets.

(c) Staurolite-phyllites.

(d) Quartzose-calcareous phyllites and conglomerates.

(a) *The calcareous phyllites.* The essential minerals are : calcite, quartz, muscovite and a basic plagioclase (andesine-labradorite) which often plays the rôle of a matrix for the other minerals. It gives a silky aspect to the rocks. The accessory constituents are : tourmaline, rutile, pyrite, epidote, zoisite, titanite, zircon, magnetite.

(b) *Grey and black phyllites with garnets.* These rocks are intercalated in the calcareous phyllites. They represent interbedded clays which have been metamorphosed. The garnets give to the rocks a knotted aspect. We also find passage-rocks between phyllites and hornfels. The essential minerals are quartz and muscovite. Calcite is rare in the phyllites containing garnet. A carbonaceous pigment gives a black colour to some of the rocks. As accessory constituents we may note : epidote, zoisite, chlorite, hornblende, plagioclase, tourmaline, zircon, rutile, apatite, magnetite, pyrite.

(c) *Staurolite-phyllites.* Phyllites with garnets may pass into phyllites containing staurolite of great dimensions.

(d) *Quartzose-calcareous phyllites and conglomerates.* In the lowest layers of the *Schistes lustrés*, we often find quartzo-calcareous phyllites as a passage-rock between the calcareous phyllites and quartzites. Conglomerates were found by Preiswerk at the basis of the *Schistes lustrés*, in the northern region of Tessin.

THE AGE OF THE SCHISTES LUSTRÉS

According to Termier's definition, the *Schistes lustrés* represent a "comprehensive series," viz. a series of sediments, 2,000–5,000 metres thick, of the same facies, but of different ages. The lower limit is not always identical. Indeed, the *Schistes lustrés* may begin in some cases at the middle Trias, in other cases at the lower Lias. They include representatives of Jurassic, Cretaceous and in some regions, Nummulitic formations.

Though they are often highly metamorphosed, the *Schistes lustrés* sometimes furnish fossils.

On the southern side of the Gothard Massif, on the Nufenen pass, J. Charpentier and Lardy, in 1814 and 1822, found *Belemnites paxillosus*. The same fossil was recorded by

Arnold Escher and B. Studer, in the vale of Piora and at the Scopi, near the Lucomagno Pass.

Since then, other fossils have been found, mostly in the *Schistes lustrés* of the Gothard group, such as *Pentacrinus tuberculatus*. *P. basaltiformis*, *P. psilonotus*, *Gryphea cymbium*, *Cardinia listeri*. Ammonites have also been recorded, but owing to metamorphism, the species was not determinable. They belong to the genus *Arietites*. We thus arrive at the conclusion that these particular *Schistes lustrés* are of Liassic age.

Franchi found several fossils, on Italian ground, in the *Schistes lustrés*, such as Ammonites (*Arietites*), Belemnites, and Corals contained in the calcareous phyllites of the valley Grana and of Narbona. He also detected Belemnites near the Lake of Verney and near the road of the Petit St. Bernard pass.

In almost non-metamorphic breccias, intercalated in the calcareous phyllites, Franchi found Belemnites at the Col de la Seigne, and Encrinus near Dronero, in the Maira valley, and near Villeneuve in the Aosta valley and Val Grisanche.

In the Upper Engadine, R. Staub has recorded in the *Schistes lustrés* of the " Margna Nappe,"—the equivalent to the north-east, in the Grisons, of the Dent Blanche Nappe—Middle Jurassic and radiolarites of the Upper Jurassic, Cretaceous (Couches rouges) and Eocene " *Flysch.*" Radiolarites have also been recorded by Termier in the *Schistes lustrés* of the Césane region, of the Mont Genèvre near Briançon, and in the region of Cairo-Montenotte. They very likely represent an abyssal facies, for they are associated with manganese ores. From a stratigraphical point of view, they indicate Middle or Upper Jurassic.

In the north-east part of the Swiss Alps, in the Prätigau valley (Grisons), Trümpy found Nummulites in the upper part of the *Schistes lustrés* (Bündner Schiefer), showing the Eocene age of this formation. Swiss geologists use the name of " *Eocene Flysch* " for this Tertiary part of the *Schistes lustrés*.

We may be astonished not to find any *Flysch* or *Schistes lustrés* of Tertiary age, in the Pennine Alps ! As foreseen by Argand and proved by Lugeon, the Tertiary of the *Schistes lustrés* of the Pennine Alps occurs *on the northern side of the*

Mont-Blanc and Aar Massifs, in the Prealps, where it has been dragged by the northward travel of higher nappes, as will be seen later. These Tertiary *Schistes lustrés* are represented by the *Niesen Flysch*.

In the Briançon zone of the French Alps we find also, as shown by Termier and Boussac, that the *Flysch* forms the upper part of the *Schistes lustrés*. The mountains of the Ubaye and of the Embrunais, situated between the Hercynian massifs of the Pelvoux and Mercantour, are mostly made up of *Flysch*, belonging to the upper part of the *Schistes lustrés* of the Briançonnais, which has been dragged towards the west and overlies the " *terres noires* " of the Autochthonous Jurassic.

To sum up, we find in the *Schistes lustrés* a lower part of Mesozoic age, represented mostly by calcareous phyllites with intercalations of greenstones, and in certain regions an upper part belonging to the Eocene, made up of shales, sandstones and breccias, representing the last phase of the filling up of the Alpine secondary geosynclines. This last part is called *Flysch*.

Thus we see that the *Schistes lustrés*, generally speaking, represent a " comprehensive series " which may include formations deposited in the Alpine secondary geosynclines from the Trias, or the Lias, up to the Eocene.

CHAPTER III

THE SIMPLON-TICINO NAPPES

Introduction

As previously seen, the Simplon-Ticino Nappes, the three lower elements of the Pennine Alps, appear at the surface of the ground owing to a culmination of pitch due to the obstacle of the Aar Massif (Fig. 51).

.The driving of the Simplon tunnel, 20 kilometres in length, which took place in the years 1898–1905, will remain for ever in the annals of Alpine geological science. This great engineering work solved the important question of the existence of nappes in the Pennine Alps at the time when many Alpine geologists did not agree with Schardt and Lugeon and criticized their interpretation of the making of mountains by huge nappes. The results arrived at are not only most important from a tectonic point of view, but also from the practical side, for the high temperatures and great springs met with furnished an instructive experience for the future.

We shall first study the geological section across the Simplon, as shown by the observations made during tunnelling, then we shall examine what is known about the temperatures and the subterranean waters in the different rocks forming the Simplon region.

Tectonics

Schardt has shown (Fig. 52) that the Simplon region is mostly made up of three nappes, each with a recognizable Paleozoic core. From top to bottom they are :

III. The Monte Leone Nappe.

II. The Lebendun Nappe.

I. The Antigorio Nappe.

The gneissic core of the *Antigorio Nappe*, representing the lowest tectonic element of the Pennine Nappes, unites with

154

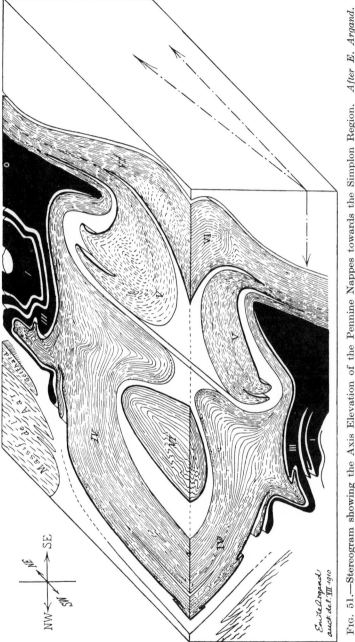

Fig. 51.—Stereogram showing the Axis Elevation of the Pennine Nappes towards the Simplon Region. *After E. Argand.*
I–III, Simplon-Ticino Nappes. IV, Great St. Bernard Nappe. V, Monte Rosa Nappe. VI, Dent Blanche Nappe.

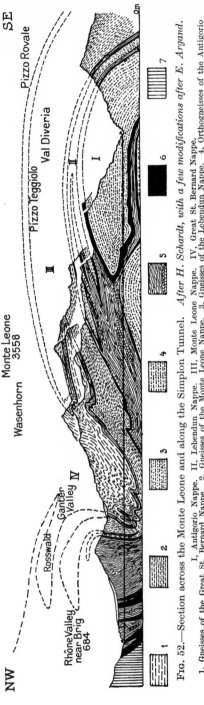

FIG. 52.—Section across the Monte Leone and along the Simplon Tunnel. *After H. Schardt, with a few modifications after E. Argand.*

1, Gneisses of the Great St. Bernard Nappe.   I, Antigorio Nappe. II, Lebendun Nappe. III, Monte Leone Nappe. IV, Great St. Bernard Nappe.   2, Gneisses of the Monte Leone Nappe. 3, Gneisses of the Lebendun Nappe. 4, Orthogneisses of the Antigorio Nappe. 5, "Schistes lustrés." 6, Trias. 7, Aar Massif.

the gneiss massif of Verampio which belongs to the Gothard Massif, itself a digitation of the Aar Massif. The Verampio Massif thus unites the lowest Pennine nappe with the Foreland. The continuation of the Verampio Massif to the northeast is represented by the gneiss of the Lucomagno in the Ticino region. The Lucomagno and Verampio Massifs are separated from the Gothard Massif by the syncline of the *Schistes lustrés* of Bedretto and Piora. R. Staub has pointed out, in his book *Der Bau der Alpen*, that, owing to the fan structure of the Gothard Massif and the petrology of both the Lucomagno and Verampio Massifs, the latter should be considered as the lowest tectonic element of the Pennine Alps, and not as belonging to the Gothard Massif. Staub's idea is a new working hypothesis.

The Antigorio Nappe. The core of the Antigorio Nappe is

made of orthogneiss. The deep gorge of the Diveria, on the southern side of the Simplon Pass, has been eroded out of it. The northern archbend of this nappe may be seen, on Italian territory, at the northern side of the Pizzo Teggiolo in the Cairasca valley. When descending from the Simplon Pass to Italy, the road enters the Antigorio gneiss 2·5 kilometres to the south-east of the village of Simplon, after having crossed *Schistes lustrés* with garnets and the Triassic dolomitic limestones of the " Alte caserne."

The Lebendun Nappe is separated from the Antigorio Nappe by a syncline of *Schistes lustrés* framed by Triassic dolomitic rocks. This syncline shows splendidly to the north of the Pizzo Teggiolo, referred to above, and hence has been called the Teggiolo syncline. The core of the Lebendun Nappe is made up of paragneisses only 250–700 metres thick. The Simplon road passes on the very thin Lebendun gneiss, one kilometre after leaving Gabi, and continues on it as far as the " Alte caserne." One can see there that the Lebendun Nappe is intercalated between two zones of *Schistes lustrés* with Trias. The upper zone supports :

The Monte Leone Nappe. Very schistose gneisses with white mica, forming the core of this upper nappe, are exposed on the left side of the road leading from the Simplon Pass to Italy. The summit of the Monte Leone represents a higher small digitation of this nappe capping a small involution of the Great St. Bernard Nappe. Owing to the obstacle formed by the Aar Massif, the frontal digitations of the Monte Leone Nappe, near the Rhone valley, are almost overturned.

At the very summit of the Cherbadung and Helsenhorn, the Monte Leone Nappe is capped by a thin band of *Schistes lustrés* supporting an outlier of the Great St. Bernard Nappe. The Ofenhorn, on the other hand, is entirely carved out of the Monte Leone Nappe.

Ascending the Simplon Pass, from Brigue, we cross a zone of *Schistes lustrés* as far as " Refuge II." Between Eisten and the Ganter bridge occur the gneisses of two vertical frontal digitations of the Monte Leone Nappe. At the Ganter bridge appears the syncline, made of schistose limestones and dolomitic Triassic rocks which separates the Monte Leone Nappe from the Great St. Bernard Nappe. At Berisal we enter the micaschists of the Great St. Bernard core, and these form the slope of the mountain as far as the Simplon Pass.   From the latter we notice that the gneissfold of the Huebschhorn, belonging to the Monte Leone Nappe, plunges underneath *Schistes lustrés*, followed by gneisses of the Great St. Bernard Nappe.

The Culmination of Pitch.   It seems that owing to the pitching towards the north-east the equivalent of the three Simplon nappes should be easily found on the left side of the Leventina valley.   Detailed mapping by Preiswerk shows that this is not the case and that the Simplon-Ticino culmination is complicated by a transversal syncline, in which appears the Great St. Bernard Nappe.   This transversal syncline occupies the left side and the upper part of the Maggia valley.   On both sides of this syncline a transversal anticline has been formed.   The Monte Leone Nappe is exposed in the anticline to the south-west of the tectonic depression, and to the north-east the gneiss of the Verzasca is seen in the other anticline.

The following gneisses take part, according to Preiswerk,[1] in the formation of the Simplon-Ticino culmination :

1. The *lower Ticino gneiss* represented by the three Simplon-Ticino nappes referred to above.   This gneiss occurs in the region limited by the following localities : Brigue, Monte Leone, Passo Selarioli, Domodossola, Bignasco, Ofenhorn, Binn.   This gneiss disappears underneath the Great St. Bernard gneiss of the transversal syncline and forms, farther to the north-east, the two sides of the Leventina valley, where it can be seen from the railway line descending to Bellinzona, between Rodi and Claro.

2. The *upper Ticino gneiss*, including the Verzasca and Simano gneiss.

The Simano gneiss is overlain by the Adula gneiss.   R. Staub and Jenny interpreted the Adula gneiss as being the equivalent of the Great St. Bernard Nappe on the north-east side of the Simplon-Ticino culmination.   But Preiswerk, who mapped in this region, pointed out that it is not possible to follow the Adula gneiss towards the south-west, and that it does not unite with the great outlier of the Great St. Bernard Nappe occurring in the Maggia valley.   In Preiswerk's opinion

[1] Preiswerk sees in the Lebendun Nappe an involution of the Great St. Bernard Nappe.   This is not the case, as pointed out by Staub, for the Lebendun gneiss roots to the south.

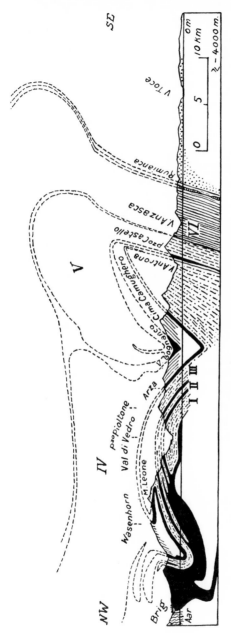

FIG. 53.—Section across the Simplon Nappes and their Roots. *After E. Argand.*

I-III, Simplon Nappes. IV, Great St. Bernard Nappe. V, Monte Rosa Nappe. VI, Dent Blanche Nappe. In black : "Schistes lustrés."

the Adula gneiss represents a higher transversal digitation of the upper Ticino gneiss.

Staub having lately maintained his interpretation, it seems at present that the equivalent of the Great St. Bernard Nappe on the north-east side of the culmination is to be found in the Adula Nappe (Fig. 49).

THE HYDROLOGY AND GEOTHERMICS OF THE SIMPLON TUNNEL

Hydrology. The strongest underground drainage met with during the driving of the Simplon tunnel was due to joints or faults. When the rocks were well-bedded, some of them being better water-bearers than others, the springs appeared at the contact between permeable and impermeable rocks. These waterways were numerous but not important. Of the different rocks of the Simplon tunnel, the Triassic dolomitic limestones with beds of gypsum, owing to their solubility, stored much more water than the gneisses, even when the latter were crushed.

The most interesting phenomena occurred between 3,800 and 4,420 metres from the southern entrance of the tunnel. The different kinds of water met with in this region have been classified by Schardt, the geologist of the Simplon tunnel, as follows :

1. *Warm waters,* all situated in the Antigorio gneiss, having a temperature higher than the temperature of the rocks.

2. *Isothermal waters* having the temperature of the rocks. Some of them contained sulphates in solution, others not, but all were more or less ferruginous.

3. *Cold waters* with a run-off 15–20 times greater than in the preceding cases.

From the southern entrance, at Iselle, no water had been met with over a distance of 3,800 metres, in the Antigorio gneiss. The first underground drainage occurred in the gneiss, much fissured, between 3,830 and 4,100 metres. From this point to the contact with the underlying Triassic limestones, the gneiss being massive was quite dry. In several places springs came out from the Triassic limestones, but the strongest drainage occurred suddenly, at 4,400 metres, on the 30th of September, 1901, and stopped the work for six weeks. In December of the same year, the tunnel crossed a fault at 4,421 metres and entered a zone of crushed micaschists containing patches of marble. This moving formation interrupted the work for several months.

The Simplon tunnel consists of two parallel tunnels, and several transverse connecting galleries permitted additional interesting observations. Indeed, at 3,860 metres, Schardt found in the same fault, and in contact, waters of different temperatures, one flowing down (25·7° C.), the other flowing up (32° C.). This important fact, confirmed by other observations in other parts of the tunnel, shows that it is not necessary that waters of different temperatures and different chemical composition should follow different underground channels. In our case we have to deal with surface water which penetrates to a great depth where it finds its passage barred by the closing of the fault. This water being heated is forced to ascend to the surface by the same fissure by which it has travelled downwards.

The cold waters are limited to the water-bearing beds of the Trias referred to above. They contain an average of 1 gramme of calcium sulphate per litre. Schardt has calculated that this underground drainage dissolves 10,000 cubic metres of calcium sulphate in a year. Water of the Cairasca torrent, coloured by fluoresceine, appeared in several cold springs in the tunnel showing a relation between surface waters in the Triassic rocks and underground drainage at a depth of 1,200 metres.

The Geothermics. In the tunnel the temperature of the rocks was studied at 200 different stations, while at the surface of the mountain it was measured at fourteen stations. The thermometers were placed in hollows 1·5 metre deep.

Before the beginning of the boring of the tunnel different forecasts had been made as to the temperature that would be recorded during the work. This was of paramount importance in order to facilitate the stay of the workmen in the tunnel. On the basis of the experience obtained in the Gothard tunnel, an engineer, Stapff, pointed out that a maximum temperature of 53° C. would be met with, for the tunnel passes 2,900 metres underneath the summit of Monte Leone. Lommel, a railway director, criticized this figure and submitted that the temperature would not exceed a maximum of 35° C., while Professor Heim calculated a maximum of 38°–39° C. *The maximum*

s.a.

recorded was 55° C., but this maximum did not correspond to the highest point of the mountain surface ; it occurred 700 metres northwards. This very high temperature was due : first, to the fact that the rocks were quite dry, and, second, to the presence of horizontal layers, caused by the geological structure, which prevented the inner heat from escaping to the surface of the ground. As soon as cold underground waters occurred, their effect made itself felt in lowering the temperature of the rocks. For instance, the Geoisotherms, or planes of equal underground temperature, fell rapidly owing to the cold water met with in the Triassic rocks underlying the Antigorio gneiss. A temperature of only 20° C. occurred in this region, where otherwise one might have expected to find 38° C.

Some of the more notable records, after Schardt, are as follows :

Stations	Altitude			Centigrade Temperature			Rate of rise metres per 1° C.	Thermal gradient per 100 metres
	Surface	Tunnel	Difference	Surface	Tunnel	Difference		
Im Rafje (Brigue) . .	686	686	0	6·3	6·3	0	—	—
Brigerberg . . .	915	688	227	8·3	16·7	8·4	27	3·7
Rosswald . . .	1,850	693	1,157	3·8	28·0	24·2	47·8	2·09
Unter Berisal . .	1,320	695	625	7·4	30·5	23·1	27·0	3·7
Hohenegg . . .	2,030	699	1,331	3·4	45·0	41·6	31·7	3·15
Furgenbaumpass .	2,690	703	1,987	−1·6	53·0	54·6	36·3	2·75
Rossetto . . .	2,221	704	1,517	1·6	49·0	47·4	32·0	3·12
AMOINCIEI . . .	2,709	693	2,116	−1·82	41·5	43·32	48·8	2·49
Lago d'Avino . .	2,237	682	1,555	0·4	41·0	40·6	38·3	2·6
Passo di Valle . .	2,448	671	1,777	1·2	32·0	30·8	57·2	1·73
Alpe di Valle . .	1,863	665	1,198	2·2	19·0	16·8	71·3	1·4
Passo Possette . .	2,248	654	1,594	1·8	32·0	30·2	52·8	1·9
Bugliaga . . .	1,316	642	674	6·6	30·0	23·4	28·8	3·47
Iselle (tunnel) . .	634	634	0	8·85	8·85	0	—	—

THE GREAT ST. BERNARD NAPPE

STRATIGRAPHY

The Casanna Schists. The Great St. Bernard Nappe is chiefly made up of *Casanna schists* which represent sandstones and clays, of Paleozoic age, more or less metamorphosed.

Wegmann has divided the Casanna schists into the following groups :

2. *The upper Casanna schists* are made up of sandstones and fine conglomerates. There is little metamorphism, and felspar is much scarcer than in the older series. On the other hand, dynamometamorphism has affected them strongly.

The greater part of this formation belongs to the Permian. Its base is of Upper Carboniferous age and the upper part may be in some cases formed by white foliated quartzites of lower Triassic age.

1. *The lower Casanna schists* show regional metamorphism and little contact metamorphism. The lower Casanna schists are characterized by intercalations of prasinites which represent metamorphosed tuffs derived from diabases and porphyries. The upper part of this formation is of Carboniferous age. Fossils never having been found in the lower part, it is not possible to ascertain its exact age, which, however, must be older than Carboniferous.

Anthracite in the Carboniferous Strata. Anthracite appears in the Carboniferous strata of the external digitations of the Great St. Bernard Nappe. This Carboniferous zone, called " zone carbonifère axiale " by the French geologists, extends from Sion, in the Rhone valley, to Savona, in Italy, through the Aosta valley, the Maurienne, the Briançonnais and the Maritime Alps.

Plant remains prove that these strata belong to the Upper

164

Carboniferous and the upper part of the Middle Carboniferous.

During the Great War the anthracite of the Great St. Bernard Nappe was exploited near the Rhone valley at Entremont, Isérable, Chandoline, Grône. The coal coming from these mines was very poor, the anthracite containing many impurities, as shown in the following figures, according to Albert Heim :

Mines	*Percentage of Impurities*
Entremont	27·6–47·6
Isérable	26·2–43·3
Chandoline	24·9–37·2
Grône	21·3–34·0

The Mesozoic Cover. The mesozoic strata covering the Casanna schists belong to the Trias and the *Schistes lustrés*.

TRIAS

The sediments of the Trias have been formed in shallow water, on the top of and round the Briançon geanticline. They are made up of clastic material (breccias), chemical deposits (gypsum and salt) and dolomitic limestones. Generally speaking, the Triassic outcrops are of a yellow hue contrasting with the grey-black colour of the *Schistes lustrés*.

In some cases, as pointed out by Wegmann, breccias make up the complete series of the Triassic rocks, as in the Herens valley, indicating deposits formed on the slopes of the geanticline. In fact, it was these shallow water deposits which led Argand to the idea of the geanticline. In the Hérens valley, Wegmann recorded two and even three zones of breccias which may perhaps correspond to a division of the Briançon geanticline into two secondary geanticlines.

It is very difficult to know exactly the age of the upper part of the breccias, as no determinable fossils have ever been found in them in the Swiss Alps. By comparison with the series of rocks of the Briançonnais across the French frontier, one may see in them the equivalent of the *Brèches du Télégraphe* of Liassic age. The pebbles of the breccias may vary from the size of a man's head to sand grains. They are generally made up of dolomitic limestones ; quartzite pebbles are not so common and schist pebbles are rare.

Schistes Lustrés

The *Schistes lustrés* of the Great St. Bernard Nappe are made of calcareous phyllites with sericite, chlorite, albite and quartz. Rutile, titanite, tourmaline are accessory. Their dark colour is due to a carbonaceous pigment.

Intercalations and sills of greenstone are common, and increase in quantity towards the south. They are represented by prasinites containing more epidote than the prasinites of the lower Casanna schists. In some cases, as in the Hérens valley, the prasinites, according to Wegmann, represent metamorphosed ashes. Near the intercalations of greenstones, epidote and zoisite occur in the *Schistes lustrés*.

When the carbonate of lime decreases, the *Schistes lustrés* are like black Carboniferous shales. By addition of quartz grains, the *Schistes lustrés* may be transformed into quartzites.

Tectonics

The Great St. Bernard Nappe is one of the dominant nappes of the Pennine Alps. It is due to the development of the Briançon geanticline in the Alpine geosyncline (page 21), and has set in motion the Simplon nappes.

The Great St. Bernard Nappe possesses several digitations, and its tectonics are characterized by four phenomena of paramount importance :

1. The backward folding of Bagnes.

2. The backward folding of the Mischabel (Plate XI, A).

3. The formation of the " worm " due to the lamination of the upper digitation and its invagination to give what are called the " lower worm " and the " upper worm" which are merely two parts of one structure.

4. The virgation.

The Backward Folding of Bagnes. When dealing with the tectonics of the Great St. Bernard Nappe we must take into consideration the influence of the obstacle, formed by the Aar Massif, to the advance of the nappe. The Aar Massif played the rôle of a dam when the Great St. Bernard Nappe was first thrust over the Foreland. The Dent Blanche Nappe (VI) was, at that time, in contact (Plate I, Fig. 9) with the Great St. Bernard Nappe, for the Monte Rosa Nappe (V) came into being later.

PLATE VI. The Weisshorn and the Backward Folding of the Mischabel.

IV, Great St. Bernard Nappe. *S*, Sedimentaries. *VI* Dent Blanche Nappe at the Weisshorn. *Photo by J. Gaberell.*

Owing to the damming effect of the Aar Massif, the Great St. Bernard Nappe could not advance any more, while on the other hand the overlying Dent Blanche Nappe continued travelling towards the north. This movement of the Dent Blanche Nappe over the Great St. Bernard Nappe formed digitations in the latter. The highest of these digitations composed of Triassic rocks (dolomitic limestones and quartzites), was crushed and laminated into lenticles. We shall have to study later what happened to this highest digitation, called the " worm," when the Monte Rosa Nappe was intercalated between the Great St. Bernard Nappe and the Dent Blanche Nappe.

The advance of the Dent Blanche Nappe and its compression forced the upper part of the digitation underlying the " worm " to fold backward (Plate I, Fig. 10). This phenomenon, called the " backward folding of Bagnes," was recorded for the first time by Argand, and studied in detail by Wegmann, one of his pupils. As shown on Fig. 54, the front part of the digitation of the Great St. Bernard Nappe, affected by the compression of the Dent Blanche Nappe, takes a fan-like structure, called also " the fan of Bagnes," for the moûntains of the Bagnes region have been carved out of it.

The Backward Folding of the Mischabel. On the left bank of the Mattervisp river, between Randa and the summit of the Weisshorn, we miss the Monte Rosa Nappe. Indeed, on the way up to the Weisshorn Hut of the Swiss Alpine Club, we cross the Casanna schists of the Great St. Bernard Nappe and find only a thin band of Trias and *Schistes lustrés* separating them, at about 3,200 metres on the eastern ridge of the Weisshorn, from the crystallines of the Dent Blanche Nappe that form the Weisshorn (Plate VI). We thus have nappe VI (Dent Blanche) resting on nappe IV (Great St. Bernard). What happened to nappe V (Monte Rosa) ?

The Monte Rosa Nappe (V) originated in the Piedmont geosyncline, between the geanticlines of Briançon and of Dolin. The nappes of Great St. Bernard and Dent Blanche being already formed, the Monte Rosa Nappe penetrated between them. During its development the Monte Rosa Nappe pushed into the back of the Great St. Bernard Nappe, producing the " backward folding of the Mischabel." The Mischabel have been carved out of this part of the Great St. Bernard Nappe.

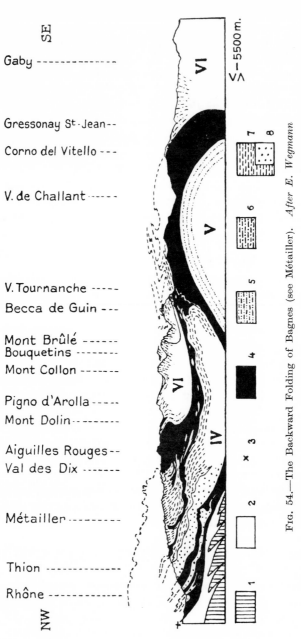

FIG. 54.—The Backward Folding of Bagnes (see Métailler). *After E. Wegmann.*

1, Crystallines of the Foreland. 2, Sedimentaries of the Foreland. 3, Thrust plane of the Foreland. 4, "Schistes lustrés." 5, Monte Rosa Nappe (V). 6, Dent Blanche Nappe (VI). 7, Great St. Bernard Nappe (IV). 8, Carboniferous of the Great St. Bernard Nappe, near the Rhone Valley.

Concurrently the highest digitation of this nappe, referred to above as the " worm," had to follow the same movement and has thus been invaginated (Fig. 55).

The Worm. The highest digitation of the Great St. Bernard Nappe, divided into several secondary digitations or slices, may be followed with clearness at Zermatt and its neighbourhood. As referred to above, this digitation has been crushed and laminated, so as to form lenticles and slices, owing to the travelling towards the north of the Dent Blanche Nappe. It is the " worm," due to the movement of the Monte Rosa Nappe plunging into the Great St. Bernard Nappe (see 4″ Plate I, Figs. 10–13). This phenomenon, to which is also due the backward folding of the Mischabel, is clearly demonstrated on the right side of the Zermatt valley, between the Riffelberg and the Bösentrift. Indeed, the greenstones and *Schistes lustrés* of the Riffelberg, representing the Jurassic cover of the Monte Rosa Nappe, dip towards the north. They are overlain by the backward folding of the Mischabel at the Bösentrift.

To the west of the Zermatt village the strong pitching towards the south-west makes itself felt, and the last outcrops of prasinites of the Mesozoic cover of the Monte Rosa Nappe die out at the southern entrance of the railway tunnel of Egg. This is the very point where the worm comes out. It is not possible to see the exact place where it unites with the backward folding of the Mischabel, the valley not being deep enough. But the great archbend not being far off (1 km.), we can imagine it. Here the worm is made up of Triassic quartzites and dolomitic limestones resting on prasinites and overlain by *Schistes lustrés.*

As a series of lenticles and slices, separated by *Schistes lustrés*, the worm rises from the tunnel towards the south. This is the " lower worm." It passes at the very foot of the Trift gorge, above the Chamois garden, and forms the base of the cliff above the villages of Herbriggen, Hubel and Zmutt. It is conspicuous from some distance (page 187) owing to its yellow hue, contrasting with the grey colour of the *Schistes lustrés* into which it penetrates. Then it appears, higher up, in lenticles, at the foot of the Hörnli, then farther to the south at the foot of the Theodulhorn. Argand has followed it to the Valtournanche, on the Italian side of the Matterhorn, and

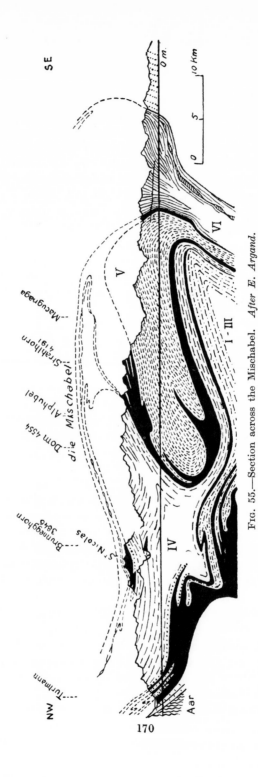

FIG. 55.—Section across the Mischabel. *After E. Argand.*

I-III, Simplon Nappes.   IV, Great St. Bernard Nappe.   V, Monte Rosa Nappe.   VI, Dent Blanche Nappe.   In black : " Schistes lustrés."

170

as far as the Alagna valley, where it bends up to form the " upper worm," passing above the localities where we noticed the " lower worm." Then we see it at the Theodulhorn, where it is separated from the " lower worm " only by a thin patch of *Schistes lustrés*. It occurs near the Swiss Alpine Club Hut at the very base of the Matterhorn. Finally, it is conspicuous along the border of the ledge of *Schistes lustrés* situated at the base of the Unter Gabelhorn, then at Trift-kummen and near the Biesjoch (3,724 m.).

Argand has shown that the worm may be followed to the south-west in the Tourtemagne valley, the Anniviers valley and as far as Evolène in the Hérens valley. Indeed, in the cliff above the Evolène village, Wegmann found many slices and lenticles made up of dolomitic breccias, dolomitic lime-stones, quartzites and breccias. The lowest of these slices, forming the " worm," overlies the *Schistes lustrés* of the Mesozoic cover of the Great St. Bernard Nappe.

To the north-east of Zermatt we see the " worm " underneath the summit of the Unter Rothhorn.

An interesting fact, well worthy of notice, is that the rocks forming the " worm " have not been touched by the intrusion of the basic magmas into the *Schistes lustrés*. Bartholmes, who studied the greenstones of the Dent Blanche region, thinks that this observation seems to demonstrate that the basic magma was intruded into the *Schistes lustrés* before the " Monte Rosa phase," viz., before the plunging of the Monte Rosa Nappe into the back of the Great St. Bernard Nappe. Professor O. T. Jones and Professor Cox, during the Alpine excursion of the Geologists' Association in 1926, under the guidance of the author, pointed out that their observations, in other regions, show that intrusions always avoid massive rocks like the dolomitic limestones and quartzites forming the worm.

The Virgation. The carapace of the Great St. Bernard Nappe shows a bundle of folds at the Pelvo d'Elva, Vanoise and Valsavaranche which represent, according to Argand, three diverging folds of a virgation characterizing the inner part of the Alpine bow.

Here we have to deal with a *confined virgation* (page 14), a virgation due to obstacles represented by some of the Her-cynian massifs of the Foreland.

The diverging fold of the Pelvo d'Elva was formed along

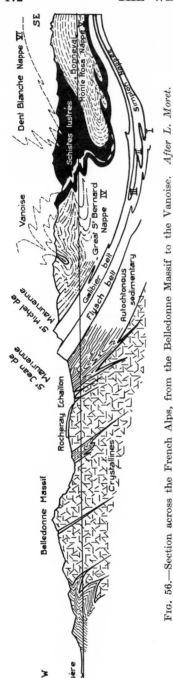

FIG. 56.—Section across the French Alps, from the Belledonne Massif to the Vanoise. *After L. Moret.*

the northern border of the Hercynian massif of the Mercantour (Argentera of Italian geologists). The diverging fold of the Vanoise is due to the effect of the Hercynian massif of the Pelvoux. The effect of the Dolent promontory, in the north-east extremity of the Mont-Blanc Massif, is marked by the diverging fold of Valsavaranche, which lengthens out in the Mischabel.

The phenomenon of the backward folding of the Mischabel and of Valsavaranche, referred to above, was facilitated, in Argand's opinion, by the formation of this diverging fold of the virgation.

OUTCROP OF THE GREAT ST. BERNARD NAPPE

The amplitude or width of the Great St. Bernard Nappe, the greatest of the Pennine Nappes, is 95 kilometres, its mean thickness being 6 kilometres. Argand pointed out that in some cases the thickness of this nappe may attain 15 kilometres.

When dealing with the Simplon-Ticino Nappes we saw the rôle played in this region by the Great St. Bernard Nappe (page 158). Between the Simplon region and the Saas valley, the Laquinhorn and Weissmies belong to the Great St. Bernard Nappe. We have seen that the Mischabel, between the Saas valley and the Zermatt valley, are also carved out of this nappe. The Turtmanntal, the greatest part of the Anniviers valley, the Réchy village, the Hérens valley as far as Evolène, the Bagnes valley, the region situated between the latter and the Dora Baltea, with the Combin de Corbassière, Mont Velan and the Great St. Bernard, belong to it.

The Great St. Bernard Nappe occurs between the Dora Baltea valley and the Isère valley in the Valsavaranche and Val Grisanche as well as at the Rutor.

Between the Isère valley and the Durance valley the same nappe appears at Mont Pourri, the Aiguille du Midi, the Bozel valley, the Pointe de la Fenêtre, the Massif de la Vanoise, between St. Michel and Modane, and at Mont Tabor.

From Briançon we follow the Great St. Bernard Nappe as far as the Mediterranean Sea, near Savona, through the Pelvo d'Elva.

CHAPTER V

THE MONTE ROSA NAPPE

Introduction

The Monte Rosa Nappe was named after Monte Rosa, the highest peak of the Swiss Alps, to be carved out of it.

This nappe is due to the development of the Dolin geanticline, or rather to the development of the Dent Blanche Nappe, and represents the bottom of the secondary geosyncline of Piedmont (Plate I, Fig. 11).

When dealing above with the tectonics of the Great St. Bernard Nappe, we realized the importance of the northward travel of the Monte Rosa Nappe. This nappe was intercalated between the nappe of the Great St. Bernard and the Dent Blanche Nappe, and produced the backward folding of the Mischabel-Valsavaranche, and the invagination of the " worm."

Rocks

As in the Simplon-Ticino Nappes, the rocks forming the Monte Rosa Nappe may be divided into two groups :

1. Paleozoic.

2. Mesozoic.

Let us study :

Paleozoic. The core of the Monte Rosa Nappe is̄ mostly made up of orthogneisses which are exposed in nearly the whole of the Monte Rosa if we except the highest ridges. Granite with tourmaline occurs at the Betemps Hut of the Swiss Alpine Club, where it forms good *roches moutonnées.*

Paragneisses surround the orthogneisses and granite. They form the higher parts of the Dufourspitze and of the Lyskamm, the whole mass of the Castor, as well as the ridge Stockhorn-Hohthäligrat.

174

The paragneisses are recognizable, owing to their reddish hue, as forming the left lateral moraine of the Findelen glacier, while the right lateral moraine, of a greenish colour, is made up of amphibolites of the Rimpfischhorn.

Mesozoic. The base of the Mesozoic series is made up of quartzites, dolomitic limestones and " corgneules " (dolomitic breccias) of Triassic age. They are overlaid by *Schistes lustrés*. The Triassic formations occur in about seven overfolded anticlines forming the upper frontal digitations of the Monte Rosa Nappe. They are conspicuous on the ridge running from the Gornergrat to the Hohthäligrat and Stockhorn. The Trias, with its yellow hue, then appears on the left side of the vale of Mattmark, on the left side of the Schwarzenberg glacier, where the " corgneules " often contain gypsum. We follow the Trias farther to the north, to the eastern margin of the Kessjen glacier, between Klein Allalin and the Egginerhorn.

The *Schistes lustrés*, with their greenstones (gabbros, prasinites, amphibolites, eclogites, serpentines), extend from Zermatt to the Saas valley, where we find them near Saas Fée and between the Weissmies and the Portjengrat, where they separate the Monte Rosa Nappe from the backward folding of the Mischabel.

Mountains are carved out of the greenstones ; for instance, the Riffelberg, Strahlhorn, Rimpfischhorn, Allalinhorn, and Fluchthorn.

To the west of the Zermatt valley, the *Schistes lustrés* and greenstones of the Monte Rosa Nappe may be followed through the Theodulpass to the Val Tournanche. Then they surround the massifs of the Gran Paradiso, Ambin and Dora Maira. Pollux, Breithorn and Klein Matterhorn are made up of greenstones.

TECTONICS

The axis elevation to the north-east allows us to see that the frontal part of the Monte Rosa Nappe is divided into two digitations by the Mesozoic syncline of the Furggtal, on the right side of the Saas valley. The upper digitation is the Portjengrat anticline, and the lower one is represented by the Latelhorn anticline.

To the south-west of the Zermatt valley, the following massifs belong to the Monte Rosa Nappe :

1. Gran Paradiso.
2. Ambin.
3. Dora Maira.

Owing to the pitching towards the north-east on the other side of the ˙Simplon-Ticino culmination, we find again the Monte Rosa Nappe in the Grisons. R. Staub, on the basis of Argand's discoveries in the Pennine Alps, pointed out that the massifs of Tambo and Suretta were equivalent to the Monte Rosa Nappe.

We have seen (page 158), when dealing with the Simplon-Ticino nappes, that Staub correlated the Great St. Bernard Nappe with the Adula Massif. Having adopted Staub's opinion, the equivalence of the Monte Rosa Nappe to the Tambo-Suretta Nappe, as mentioned above, seems well demonstrated.

With the latter, the Monte Rosa Nappe seems to disappear for ever, for a new series of nappes occurs in the Upper Engadine, overriding the Pennine elements. These nappes are *the Austrides* of R. Staub.

REAPPEARANCE OF THE MONTE ROSA NAPPE

A glance at the tectonic map (Plate XII) shows that the *Schistes lustrés* reappear among the Austrides in the *Hohe Tauern window*. As we know that the *Schistes lustrés* are characteristic of the Pennine Nappes, we can agree, with P. Termier's conclusions that the Hohe Tauern window proves that *the Pennine Nappes do not die out in the Grisons, with the Suretta and Margna Nappes, but that they extend underneath the Austrides.*

Two crystalline massifs come out of the *Schistes lustrés* of the Hohe Tauern window :

1. The massif of the Venediger, at the west.
2. The massif of the Hochalm, at the east.

The results arrived at by Termier, in 1903, were developed by Kober who has shown that in the crystalline massifs of the Hohe Tauern window we have to deal with four Pennine nappes. These piled-up nappes are, from the top to the bottom :

4. The Modereck Nappe.

3. The Sonnblick Nappe.

2. The Hochalm Nappe.

1. The Ankogel Nappe.

R. Staub demonstrated that the nappes of the Ankogel and Hochalm are the equivalent of the Monte Rosa Nappe. As we shall see later, R. Staub correlates the higher nappes, Sonnblick and Modereck with the Dent Blanche Nappe (Plate XI, c).

S.A.

CHAPTER VI

THE DENT BLANCHE NAPPE

INTRODUCTION

The Dent Blanche Nappe (VI) is due to the development of the Dolin geanticline (see Plate I). It is the highest tectonic element of the Pennine Nappes. Thus it is impossible to find it on axis culminations, for it has been removed by erosion. But we find it, well developed, in the transversal syncline, situated between Monte Rosa and the Gran Paradiso. There the Dent Blanche Nappe floats on its substratum of *Schistes lustrés* (Fig. 50 and tectonic map, Fig. 61).

Owing to the backward folding of the Mischabel (see page 167), the Dent Blanche Nappe comes into contact with nappe IV (Great St. Bernard) and not with nappe V (Monte Rosa).

PALEOZOIC

The Paleozoic rocks of the Dent Blanche Nappe have been classified by Argand, as follows, from the top to the bottom :

2. Arolla series.

1. Valpelline series.

Each series is made up of : (*a*) rocks of eruptive type ; (*b*) paragneisses representing sedimentary formations which have been metamorphosed ; (*c*) passage-rocks between these two types.

The characteristics of each series are as follows :

2. Arolla series, 3,000–4,000 m. thick.

{ Quartzites and chlorite-sericite-micaschists, and fine chlorite-albite-gneisses, often with epidote.—*Sedimentary.*
Injection gneisses.—*Passage-rock.*
Granites and orthogneisses.—*Eruptive.*

178

1. Valpelline series.

$\Big\{$ Micaschists and fine gneisses containing much graphite, with intercalations of calcareous strata, including marble ;   biotite-gneisses ; sillimanite - gneisses,   kinzigite - gneisses.[1]— *Sedimentary.*

Injection gneisses.—*Passage-rock.*

Granites and orthogneisses ;   diorites ;   gabbros and peridotites.—*Eruptive series.*

The Arolla Series, as clearly shown by Argand, is well exposed on the left side of the Zinal glacier, in the great cliff of the Bouquetins and of the Pigne de l'Allée.   Indeed, the upper part of the scarp is made up of granite of a light hue, overlying the crystalline schists, of a darker colour.   The contact between the two formations is very irregular.

From a tectonic point of view, it is advantageous to consider the Arolla series as a unit.

The upper part of the Arolla series passing gradually, in many cases, to the *Schistes lustrés*, it is not possible to determine its age exactly.   It may belong to the lowest Mesozoic or to the upper Paleozoic.

The tectonic relations between the Valpelline and Arolla series enabled us to fix the stratigraphical sequence of the Paleozoic rocks of the Dent Blanche Nappe.

It is worth noting that the granite batholiths of the Dent Blanche Nappe, owing to the recumbent folds, are upside down, with the granite on the top of the gneisses.

The Valpelline Series, in the scenery, is easily distinguishable from the Arolla series, owing to its darker hue.   For instance, it is not necessary to be a geologist to notice that the dark coloured terminal pyramid of the Matterhorn, above the Swiss shoulder and the Tyndall ridge, is made up of other rocks than the great grey cliff of the mountain.   Indeed, the terminal pyramid of the Matterhorn belongs to the Valpelline series, as shown by Novarese, and the Arolla series forms the great underlying scarp.

In Argand's opinion it is not possible to find a well-marked contact between the Valpelline and Arolla series.   The Valpelline series passes gradually into the Arolla series and does not form an independent nappe.

[1] Gneisses with garnet, sillimanite, andalusite, cordierite and graphite.

Mesozoic

The existence, at Mont Dolin (left side of the Arolla valley), of remnants of the normal series of the sedimentary cover of the Dent Blanche Nappe is of paramount importance. As shown by Argand, they enable us to understand that the Dent Blanche Nappe was born from a geanticline of the great Alpine geosyncline.

Haug has recently called in question the existence of the Dolin geanticline, thus contesting the validity of Argand's interpretation of the formation of the Pennine Alps. Those who have been to the Dolin have been struck with wonder at the splendid evidence given by the facts, as was the case with the English geologists who accompanied me to the Dolin in 1924. Indeed, the Dolin is one of the keys of Argand's synthesis. One has to go and see the breccias made up of huge blocks of quartzites and dolomitic limestones to realize that the top of the geanticline was demolished by wave action. Moreover, the presence of dolomitic limestones, dolomitic breccias, grey limestones and quartzites, point to shallow water deposits. The evidence of the existence of a geanticline is even more clearly demonstrated than in the case of the Great St. Bernard Nappe. These shallow water deposits represent the Trias and the lower part of the Jurassic.

If, on the other hand, we study the reversed limb of the Dent Blanche Nappe, along the left side of the Zmutt glacier we see that shallow water deposits do not occur between the Arolla series and the *Schistes lustrés*. We are here in the deeper part of the Piedmont geosyncline, and the lowest part of the *Schistes lustrés* represents the bathyal facies of the Trias.

Tectonics

That the Dent Blanche Nappe once extended as far as the Mediterranean Sea was demonstrated by R. Staub, who sees remnants of the Dent Blanche Nappe in the outliers of Mont Genèvre, Col d'Apet, Col de l'Eychauda, near Briançon, and in the crystalline of the Savona region. In R. Staub's opinion the presence of radiolarites in the *Schistes lustrés* of these localities demonstrates that they belong to the Margna Nappe and hence to the Dent Blanche Nappe.

From a tectonic point of view, according to Argand, the

Dent Blanche Nappe contains three digitations, as follows, from the top to the bottom (Fig. 57) :

3. Dent Blanche, in restricted sense.
2. Mont Mary. ⎱ In the Aosta valley.
1. Mont Emilius. ⎰

THE DENT BLANCHE NAPPE IN THE GRISONS

We have seen above (page 146) that, owing to the pitching towards the north-east from the Simplon-Ticino culmination, the Great St. Bernard and Monte Rosa Nappes reappear in the Grisons. As shown by R. Staub, the Dent Blanche Nappe has its equivalent in the *Margna*, a mountain on the right side of the Lake of Sils, in the Upper Engadine.

R. Staub ascertained that the *Maloggia series* is nothing but the Arolla series, and that the *Fedoz series* represents the Valpelline series. Thus, we arrive at the following conclusions :

Pennine Alps			*Grisons*
Dent Blanche Nappe	.	.	Margna Nappe.
2. Arolla series	.	.	2. Maloggia series.
1. Valpelline series .	.	.	1. Fedoz series.

R. Staub pointed out that the Sella Nappe, which was considered before as a well-defined nappe, was only a higher digitation of the Margna Nappe.

The sedimentary cover of the Margna Nappe contains *Flysch*, which extends from Oberhalbstein to Prätigau. In Staub's opinion, the Margna Nappe is the only one of the Pennine nappes containing *Schistes lustrés* with *Flysch*. This statement supposes that the *Niesen Flysch*, in the Prealps, has to be correlated with the Dent Blanche Nappe, and not with the Great St. Bernard Nappe, as postulated by Lugeon. We shall discuss this point when dealing with the Prealps (page 239).

The Dent Blanche-Margna Nappe was crushed and laminated owing to the overriding of the Austrides. The crystallines of the lowest digitation stop at the Septimer Pass, while detached slices have been piled up to the north. They represent the zone of the Schams slices (schuppenzone), which unites with the Oberhalbstein *Flysch*.

The highest part of the Margna Nappe, and hence the

highest part of the Pennine Nappes is formed by *Schistes*

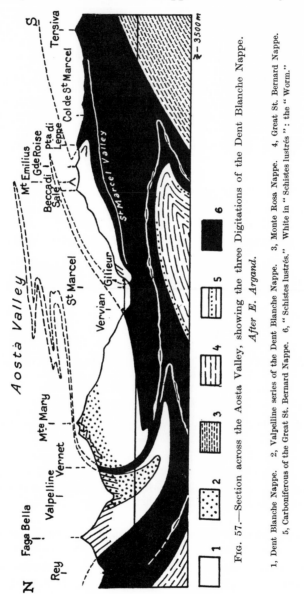

FIG. 57.—Section across the Aosta Valley, showing the three Digitations of the Dent Blanche Nappe. *After E. Argand.*

1, Dent Blanche Nappe.   2, Valpelline series of the Dent Blanche Nappe.   3, Monte Rosa Nappe.   4, Great St. Bernard Nappe. 5, Carboniferous of the Great St. Bernard Nappe.   6, "Schistes lustrés."   White in "Schistes lustrés"; the "Worm."

lustrés with intercalated slices of greenstones.   The Piz Platta has been carved out of them.

To sum up, R. Staub sees in the Margna Nappe the following tectonic elements : (1) A crystalline core divided into three digitations, as was the case for the Dent Blanche Nappe in the Aosta valley ; (2) the zone of the Schams slices uniting with the Flysch ; (3) an upper zone of *Schistes lustrés* with greenstones, called the Platta zone.

REAPPEARANCE OF THE DENT BLANCHE NAPPE

The Dent Blanche Nappe reappears in the two following windows of the Austrides :

The Lower Engadine window.

The Hohe Tauern window.

In the *Lower Engadine window* we recognize the base of the

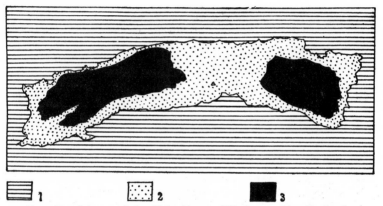

FIG. 58.—Scheme of the Hohe Tauern Window. *After R. Staub.*

1, Austrides.   2, Gross Glockner Nappe (Dent Blanche Nappe).   3, Venediger Massif (Monte Rosa Nappe).

Flysch of the Margna Nappe and the zone of the Platta greenstones.

Let us now consider the Dent Blanche Nappe in the *Hohe Tauern window* (Fig. 58).

We have seen (page 176) that Kober discovered in the Hohe Tauern window four tectonic elements belonging to the Pennine Nappes. The lower elements, the Ankogel and Hochalm Nappes, have been correlated by R. Staub with the Monte Rosa Nappe. The Sonnblick and Modereck Nappes represent the Dent Blanche Nappe. If we consider both Kober's and Staub's classification of the Pennine Nappes of the window we arrive at the following conclusions :

KOBER.   R. STAUB.

4. Modereck Nappe ⎱ Gross Glockner Nappe = Dent Blanche
3. Sonnblick Nappe ⎰   Nappe.
2. Hochalm Nappe ⎫
1. Ankogel Nappe ⎬ Venediger Massif = Monte Rosa Nappe.

As shown by R. Staub, the Glockner Nappe is characterized by slices in its upper and frontal part, as was the case with the Margna Nappe. The "Matreierzug," of the Austrian geologists, has a Pennine character; R. Staub sees in it the equivalent of the Schams slices and of the Platta zone. The "Matreierzug" occurs all round the window, at the base of the Austrides.

We have seen that *Flysch* occurs in the Margna Nappe and unites with the Schams slices. R. Staub ascertained the existence of *Flysch* in the north-west part of the Hohe Tauern window, at the base of the Tarntalerköpfe. This Tarntal *Flysch* stands in the relation to the Matreierzug, as the *Flysch* of the Grisons to the Schams slices.

R. Staub arrives at this result, that the *Flysch* of Prätigau, Arblatsch, Tarntal belong to the same series, viz., to the *Flysch* of the Dent Blanche Nappe.

THE FAR DISTANT RESISTANCE OF MONT-BLANC

As shown by Argand, in the western part of the Dent Blanche region the Mont-Blanc Massif has hindered, by its resistance, the progress of the nappe towards the north-west. The surface of contact with the substratum rises rapidly to the vertical (Veisivi) and then starts forward again (Roussette), becoming intricately digitated (Couronne de Bréonna). This disposition may be either the result of the stoppage forced upon an enormous frontal fold, as at Veisivi, or an attempt of the nappe to scale the obstacle in front of it. On the front of the moving mass, the *Schistes lustrés* of the Great Combin zone were hampered in their outflow and have accumulated to an enormous thickness, sometimes several kilometres, as in the Hérens valley.

If we walk towards the east, we see the surface of contact leaning over more strongly towards the outside, as seen at the Alpe and Col de l'Allée. The nappe now moves more freely, as proved by the lessening thickness of the Mesozoic substratum and the direction of movement shown by the

Schistes lustrés and the greenstones, as at the Garde de Bordon (Plate VII).

To the east of the Zinal valley the conditions of outflow change very rapidly. The advance of the nappe is no longer stopped, it spreads over strongly to the north-west, as at the Diablons, and carries away the Mesozoic formations of its substratum. The latter, which were so bulky in the western region, are here reduced to 200–300 metres and even less, as is evident at the Frilihorn, Brunegghorn, Weisshorn. Such a lamination is an immediate consequence of the easy motion of the nappe, and its result is the lowering of the tectonic level of the thrust plane.

The advance of the nappe is then greater to the north of the Diablons than on any other point of its frontal border. The western region, on the contrary, hampered in its outflow by the far distant resistance of the obstacle, has been left many kilometres behind.

If we sum up, we have to deal with :

1. A south-western region, extending from the Zinal valley to the surroundings of the Aosta valley, which is under the influence of the obstacle represented by the Hercynian massif of Mont-Blanc.

2. A north-eastern region which extends, at present, only to a small area of the region lying to the east of the Zinal valley (Diablons and Bieshorn).

In fact, as clearly demonstrated by Argand, a straight line drawn perpendicular to the root of the nappe and passing at the northern extremity of the Mont-Blanc Massif, divides the Dent Blanche Massif into two parts corresponding roughly to the two regions just referred to above. Their common border line runs almost along the Zinal valley.

The advance of the second region corresponds to the zone of easy motion situated between the Mont-Blanc Massif and the Aar Massif. The complete destruction of the nappe, by erosion, to the east of the Zermatt valley does not allow the exact reconstruction of this protuberance.

Argand has compared the effect of the Hercynian massifs of the Foreland on the advance of the nappes, to the effect of the piles of a bridge on the flow of a river. Between the piles the water flows rapidly, behind them it accumulates.

CHAPTER VII

THE GEOLOGY OF THE ZERMATT REGION

THE GORNERGRAT

From no other view-point in the Western Alps is it possible to command, at one glance and with such clearness, the most intricate and therefore also the most interesting relations of the three upper Pennine Nappes to one another.

The Gornergrat being the point from which the most magnificent geological panorama is to be seen, let us begin our observations there.

Through the northern gap of the Zermatt valley we see the Bietschhorn belonging to the Hercynian Foreland. Nearer, we have the Great St. Bernard Nappe with its core represented by the orthogneisses of Randa, which we find again at a greater height in the Mischabel, thanks to a tremendous axis elevation. They are surrounded by the paragneisses which form the great archbend called the "backward folding of the Mischabel" (page 167). The upper quadrant of this fold is easily detected on the left bank of the Viège, where its simple and powerful curve comes down from the Mettelhorn to the thalweg. The lower quadrant appears on the right bank, on account of the axis elevation, and passes on to the northern ridge of the Bösentrift, the Obere Täschalp, the Rothengrat and the Alphubeljoch, to go from there over into the Saas valley.

From the Alphubeljoch to Stalden the whole group of the Mischabel belongs to the Great St. Bernard Nappe.

Behind us we find the Monte Rosa Nappe. Its anticline core is represented by the mountains extending from the Schwarzenberg-Weisstor to the Zwillingpass. Cima di Jazzi, Monte Rosa, Lyskamm, Castor and the ridge Stockhorn-Hohthäligrat, between the Gorner and Findelen glaciers, all belong to that same nappe. Nearly the whole of Monte Rosa

186

is made up of the orthogneiss core which is also to be seen at the foot of the Lyskamm. The higher parts of the Dufourspitze and of the Lyskamm, the whole mass of the Castor, as well as the Stockhorn-Hohthäligrat belong to the external cover of paragneisses.

The basal part of the Mesozoic sedimentary cover, which is made of quartzites and dolomitic limestones, forms about seven overfolded anticlines easily seen on the ridge that runs from the Gornergrat to the Hohthäligrat and Stockhorn. The folds represent the upper frontal digitations of the Monte Rosa Nappe. The remainder of the Mesozoic sedimentary cover is represented by the *Schistes lustrés*. They stretch out in a narrow band from the Zwillingpass to the Schwarzenberg-Weisstor, passing over the very summit of the Gornergrat. On the top of these thin strata we find a bulky mass of greenstones which contains gabbros, prasinites, amphibolites, eclogites and serpentines. Pollux, Breithorn, Klein-Matterhorn, the base of the Theodulhorn, Lychenbretter, the slopes surrounding the Schwarzsee, the slopes of the Staffelwald, Zmutt, Hubel, Herbriggen, and Bodmen, the different outcrops in Zermatt itself, the Gorner gorges, the whole of the Riffelberg, the Strahlhorn, Rimpfischhorn and Allalinhorn are all built up of these greenstones.

To the west we notice a well-marked ledge, mostly covered by glaciers, extending from the Theodulpass to the Hörnli, at the very foot of the Matterhorn. Then our eyes follow this ledge from the base of the Unter Gabelhorn as far as to the Mettelhorn. We have to deal again with soft calcareous schists with greenstones, of Jurassic age, the *Schistes lustrés*. These *Schistes lustrés* always represent, in the Pennine Alps, the synclines between the great recumbent anticlines. The Matterhorn, Dent Blanche, Ober Gabelhorn, Rothorn and Weisshorn rest on these *Schistes lustrés*. They are all made of crystalline rocks of Paleozoic age. All these summits represent the core of the Dent Blanche Nappe, the highest tectonic element of the Pennine Alps.

Argand's " *faisceau vermiculaire* " or " worm " is to be seen in the *Schistes lustrés* that form the substratum of the Dent Blanche Nappe. It is the highest digitation of the Grand St. Bernard Nappe that has been laminated and invaginated owing to the push of the Monte Rosa Nappe. The theoretical

explanation is given on page 169. The lower part of the
"worm" appears as a series of Triassic lenticles, of a yellow
colour, starting from the railway tunnel near Zermatt. It
passes 200 metres above Zermatt, and forms the base of the
cliff above the villages of Herbriggen, Hubel and Zmutt. We
see it again at the Hörnli, at the base of the Furggrat and in
the Theodulhorn. We do not see it farther south, but it has
been followed by Argand as far as the valley of Alagna.
The upper part of the "worm" is to be seen at the Theodulhorn, where it is separated from the lower only by a thin
patch of *Schistes lustrés*. Then we notice it at the Furggrat
and near the Swiss Alpine Club Hut at the Matterhorn foot ;
finally, it is obvious on the border of the ledge situated at the
base of the Unter Gabelhorn and at Triftkummen and near
the Biesjoch (3,724 m.).

Another feature of paramount importance is exhibited at
the base of the Matterhorn cliff and better at the base of the
Unter Gabelhorn. It is a sill of gabbro marking the reversed
limb of the Dent Blanche Nappe. The intrusion of this basic
magma at this very place is interesting from the point of
view of Wegener's theory (see page 26).

From Zermatt to the Schönbuhl Hut

Geology of the Matterhorn and of the Dent Blanche Nappe.
Leaving Zermatt, we ascend to Zmutt and Kalbermatten
through the greenstones belonging to the substratum of the
Dent Blanche Nappe. Above Zmutt, at the base of the
Hohlicht cliff, we notice the lower series of outcrops belonging
to the "worm," recognizable by their yellowish colour. Above
Kalbermatten appears the upper series of the "worm," very
thick owing to a fold in the form of an "S." We soon detect,
on and in the *Schistes lustrés*, the gabbros belonging to the sill
we saw the previous day from the Gornergrat. Before climbing
the Zmutt lateral moraine we traverse a batholith upside down,
with the granite on the top of the gneisses.

Argand, it will be remembered, has determined two series
in the crystalline rocks of the Dent Blanche Nappe : *the
Valpelline series* is the older, and the *Arolla series* the younger.
The Valpelline series is formed of marbles with minerals
characteristic of contact metamorphism ; diorites, gabbros
and periodotites ; euphotides with gabbros ; granite with

PLATE VIII. The Matterhorn seen from the Hérens Pass.
Ar, Arolla series. *Vp*, Valpelline series. *Photo by J. Gaberell.*

muscovite and kinzigite-gneiss.[1]   The Arolla series is chiefly made of granite with amphibole ; orthogneisses with chlorite and sericite called " Arolla gneiss." In the scenery the Valpelline series is of a much darker colour than the Arolla series and we shall be able to recognize it from a distance.

In the neighbourhood of the Swiss Alpine Club Hut occur the marbles and gabbros of the Valpelline series. Specimens of kinzigite-gneiss may be found in the debris.

The Valpelline series forms the core of the digitations of the Dent Blanche Nappe.   It is obvious that the terminal pyramid of the Matterhorn belongs to the Valpelline series.   The great cliff underneath it, with a splendid recumbent syncline, is made of rocks of the Arolla series.   Owing to the pitching towards the south-west, the Valpelline series of the Matterhorn summit forms the top of the Stockjé.   The frontal part of the recumbent anticline corresponding to this Valpelline series is to be found above the Hut.

A great recumbent syncline is well marked in the Arolla series of the Wandfluh.   This syncline connects the anticline referred to above with a higher anticline of the Valpelline series occurring in the Tête de Valpelline.   These tectonic elements appear also in the northern cliff of the Dent d'Hérens. From the base to the summit of this peak we see : (1) Arolla series, (2) Valpelline series of the Matterhorn summit, (3) Arolla series of the Wandfluh, (4) Valpelline series of the Tête de Valpelline (Plate VII).

THE UNTER GABELHORN

The ascent of the Unter Gabelhorn, which should be undertaken in good weather, enables us to make interesting observations on the contact of the Dent Blanche Nappe with its substratum of *Schistes lustrés*.

Starting from Zermatt, we cross, below the Chamois Garden, outcrops of amphibolites with garnets, belonging to the Mesozoic cover of the Monte Rosa Nappe.   Above the Garden, *Schistes lustrés* containing prasinites are met with.

At the foot of the Trift gorge, on the left side of the torrent, the Triassic rocks (crushed quartzites and dolomitic limestones) of the " lower worm " are well exposed.   They form several

[1] Gneisses with garnet, sillimanite, andalusite, cordierite and graphite.

slices, separated by *Schistes lustrés*, easily recognizable owing to their greyish-yellow hue. Good exposures of *Schistes lustrés* can be seen on the path, a few minutes before reaching the Hotel Edelweiss. From there, the path follows the *Schistes lustrés* that, to the south-west, form a ledge above the cliff of harder rocks belonging to the " lower worm." The wide gully up which we walk to reach Hühnerknubel, at the foot of the south-east ridge of the Unter Gabelhorn, enables us to see, on the grass slope, two bands of prasinites intercalated in the *Schistes lustrés*. To the right of the path an outcrop of diorites is exposed.

At the summit of the gully, patches of greyish-yellow rocks occur in the *Schistes lustrés*. They are made up of Triassic quartzites and dolomitic limestones which represent the lowest slices of the " upper worm." Within a few minutes we reach the great ledge which, extending from the foot of the Matterhorn to the base of the Weisshorn, is due to weathering of the soft *Schistes lustrés* of the substratum of the Dent Blanche Nappe. The border of the ledge is made up of greyish-yellow dolomitic limestones representing the upper part of the " upper worm."

From here, the most striking feature in the scenery is the axis elevation of the Great St. Bernard Nappe. Looking to the north-east, we see a deep and narrow gully, eroded out of dolomitic limestones, below the Bösentrift. It marks the limit of the backward folding of the Paleozoic core of the Great St. Bernard Nappe. The different peaks of the Mischabel have been carved out of this core. Orthogneisses of the core occur in the upper parts of the Täschhorn and Dom, and are distinguishable from the dark paragneisses owing to their grey hue.

The geology of the Unter Rothorn is also interesting. Indeed, we notice, on its southern ridge, slices of quartzites and dolomitic limestones belonging to the " worm." From here one gets a better impression of the plunging of the Monte Rosa Nappe into the invaginated upper surface of the Great St. Bernard Nappe, than from the Gornergrat.

The shoulder of the Riffelberg, in front of us, and also that of the Hohlicht, where we stand at present, are, according to Argand, remnants of the bottom of an old senile valley of Pliocene age, of a width of 6–7 kilometres. These shoulders demonstrate that an uplift of about 2,500 metres took place at the beginning of the Quaternary. The present Zermatt valley is thus a "rejuvenated valley."

Resuming our ascent, we again traverse *Schistes lustrés* to reach the lowest part of the ridge of the Unter Gabelhorn. Then we soon meet with a band of Arolla gneisses. R. Staub in his book *Der Bau der Alpen* expresses the opinion that these gneisses represent the Monte Mary, or middle digitation of the Dent Blanche Nappe. This was a working hypothesis, put forward by Argand when leading an excursion of the Swiss Geological Society. Later, however, Argand arrived at the conclusion that we have to deal with a complication of the highest digitation of the Dent Blanche Nappe. If we had here the equivalent of the Monte Mary digitation, we should find the "upper worm" above as well as underneath it, which is not the case. On these orthogneisses we find *Schistes lustrés* with prasinites, overlain by a *sill of gabbro*.

Indeed, this observation represents the most interesting part of the geological excursion to the Unter Gabelhorn. This sill of gabbro, 80–100 metres thick, is well exposed at the foot of the steep part of the ascent, *on the thrust plane of the Dent Blanche Nappe*. It may be followed, at the base of the nappe, to the south-west, as well as to the north. It shows, as pointed out by Argand, that a basic magma was intruded along the thrust plane of the nappe. *Schistes lustrés*, not completely digested, accompany these gabbros.

From the point of view of Wegener's theory, these gabbros represent "Sima" that has been injected into the reversed laminated limb of the geanticline. We have seen (page 26) that the results arrived at by Argand, in the Pennine Alps, may be regarded as giving great support to Wegener's hypothesis of drifting of continents.

Resuming our ascent, we enter Paleozoic paragneisses of the Arolla series, forming the reversed limb of the Dent Blanche Nappe. Then we traverse white gabbros, of the Arolla series, containing beautiful green mica, fuchsite. The paragneisses are capped by granite with *horizontal aplite veins*, forming the upper part of the Unter Gabelhorn.

This excursion makes a very good complement to the other geological excursions in the Zermatt region.

CHAPTER VIII

THE GEOLOGY OF THE MALOGGIA REGION

THE MARGNA

The Margna (3,162 m.) is the formidable bastion which forms the south-east flank of the Maloggia Pass. It is the mountain of the lake of Sils, near which begins the north-eastwardly extended outcrop of the Austrides. To the south-west rise the mountains of the Bregaglia, with their proud Tertiary granitic peaks, which stand favourable comparison with the much older granitic "aiguilles" of the Mont-Blanc Massif.

The Margna will remain for ever in the annals of Alpine geological research, for R. Staub, following the work of Argand in the Pennine Alps, found in it the tectonics of the Matterhorn. Previously he had recognized the Arolla gneiss in the Maloggia rocks, the Valpelline series in the Val Fedoz, and lastly the equivalent of the Dolin (Arolla) breccias in the breccias of the Fextal.

Thus the Margna Nappe is nothing but the equivalent of the Dent Blanche Nappe, and *the Margna is geologically the homologue of the Matterhorn* (Fig. 59).

THE LAKES OF THE UPPER ENGADINE AND THE GEOGRAPHICAL PARADOX OF THE MALOGGIA PASS

The lakes of the Upper Engadine : St. Moritz, Campfer, Silvaplana, and Sils, constituted but one single water-sheet after the retreat of the glaciers. This was divided into a series of lakes by the deltas due to the alluvia carried by the affluent streams. One can actually witness the progression of the delta of the Val Fedoz torrent into the lake of Sils, which will eventually be transformed into two lakes.

Anyone travelling up the valley is struck by the total lack of a catchment basin above the lake of Sils. The valley begins, in fact, suddenly at the Maloggia Pass, where the lake begins as well. There is there a geographical paradox, cleverly demonstrated, years ago, by Albert Heim. The Upper Engadine lacks the whole of its higher portion, the portion which justifies the existence of a valley. The Inn valley is therefore "beheaded." The Inn, which flows into the lake of Sils, is a small rivulet instead of being a considerable torrent.

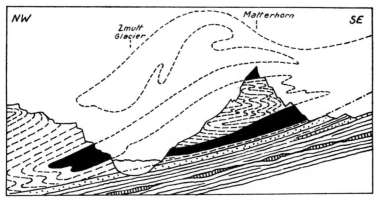

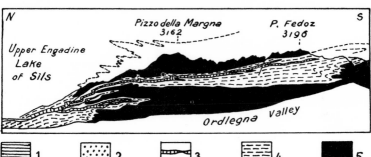

FIG. 59.—Section across the Matterhorn, *after E. Argand*, and across the Pizzo della Margna, *after R. Staub*.

1, "Schistes lustrés." 2, Greenstones. 3, Trias. 4, Arolla-Maloggia gneisses. 5, Valpelline Fédoz gneisses.

From the summit of the Maloggia one looks down a deep valley which opens south-westwards : it is the Bregaglia valley, with the Maira river. There is a problem as to the source of this stream. It is not at the foot of the wall which rises above Casaccia, and also bears the Maloggia. The Maira has, in fact, its origin in both the valleys of Maroz and Muretto, that is to say, the torrent of Val Maroz and the Ordlegna,

S.A.

which flows in Val Muretto, assume the name of Maira after
their confluence below Casaccia. The subject of the direction
of these two torrents leads to the following considerations :
the Val Maroz torrent, like the Ordlegna, flows at first towards
the north-east, then bends twice, at right angles, so as to take
the opposite direction, that is south-west, which is that of
Val Bregaglia. These bends are " bends of capture," and
the conclusion reached is that at the Maloggia there exists a
splendid example of a capture : that of the Inn by the Maria.

The upper part of the Val Maroz torrent, with its north-east
trend, represents without doubt the upper reaches of the Inn
river. The portion of the Ordlegna with a north-east direction
was an affluent of the lake of Sils. At present, from Piz
Lunghino one discovers readily the old track of the Ordlegna,
marked by the rocky terrace which from the Maloggia follows
the direction of Val Muretto. From Piz Salacina it can be
seen that Val Maroz once reached as far as the Maloggia.

In Switzerland the Inn river, the gradient of which is at
present only 8 in one thousand, was easily beheaded by the
Maira, with its gradient of 6 in a hundred.

CHAPTER IX

THE BREGAGLIA VALLEY AND ITS GRANITIC MASSIF

One of the oldest of Alpine routes, dating back to Roman times, passes through the Val Bregaglia and the Septimer, and, during the Middle Ages, it was the main line of transit between Germany and Italy. A section of this great road, paved with large stones, is still in existence above Casaccia. Val Bregaglia was also at one time used by tourists going to the Grisons from the south. It was therefore for long an important route, but the Bernina line has now ·dealt it a heavy blow.

The Bregaglia is a deep valley, with a drop of 1,100 metres within a distance of only 18 kilometres. The two sides of the upper part of the valley, between Casaccia and Vicosoprano, exhibit great morphological dissimilarity. The left side is characterized by the huge granitic massif of the Bregaglia with the Bacun, the Cengalo (3,374 m.), the Badile (3,311 m.), the Pizzi di Sciora, etc. On the other hand the right side shows the much smoother scenery of the *Schistes lustrés* with their greenstones, which form the base of the Margna Nappe and the sedimentary cover of the Suretta Nappe. It is the chain which stretches from Piz Lizun to Piz Duan (3,140 m.).

The question of the significance of the granitic massif of the Bregaglia arises naturally, since it has been shown that the Margna and the region of the Maloggia belong to the Dent Blanche Nappe. One would expect to meet in succession the deeper nappes when descending from the Maloggia to Castasegna, near the Italian frontier. This is what really takes place, especially on the right side, in the bottom of the valley and in the basal slopes of the left side. The presence of a granitic massif on the left side is due to the fact that the Pennine Nappes, after assuming their position, have been traversed by deep-seated magmas. The magmas have digested

the rocks of the nappes into which they were injected. The
final result of this great eruptive phenomenon is the granitic
massif of the Bregaglia. Thus, with R. Staub, the conclusion
is reached that the granites are younger than the emplacement
of the Pennine Nappes. Since these nappes moved during
Oligocene times, the granites are referred to the Miocene
(Albert Heim) (Fig. 60).

The granitic massif of the Bregaglia is of interest to the

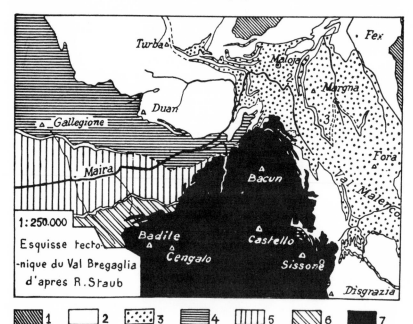

FIG. 60.—Tectonic Map of the Bregaglia Valley. *After R. Staub.*

1, Crystallines of the Austrides. 2, Mesozoic of the Pennine Nappes. 3, Margna Nappe.
4, Suretta Nappe. 5, Tambo Nappe. 6, Adula Nappe. 7, Granitic Massif of the Bregaglia.

geologist as well as to the Alpine climber. For the geologist
the best points are the contact zones between the granite and
the nappes, where a variety of well-marked contact phenomena
is to be seen. Thus the region of the Passo di Vazzeda, not
far from the Forno Hut, has been famous since the publication
of one of Staub's works. There aplite veins are to be seen,
traversing Triassic dolomitic limestones.

The granitic massif of the Bregaglia is geologically very
different from that of Mont-Blanc. The latter, much older

PLATE IX. Pizzi di Sciora (Bregaglia Valley), Consisting of Tertiary Granite.

Photo by J. Gaberell.

and of Permian age, had to bear the full wave-like impact of the Pennine Nappes. Since it is rigid, there could not be any folding, and in consequence it broke into huge overlapping wedges ! Such a structure is non-existent in the Bregaglia, because, as shown, the nappes had already come to rest before the granitic magma pierced them (Plate IX).

Numerous landslips in the nappes are a striking feature met with when going up the Val Bregaglia. They are seen on both sides of the valley from Castasegna to Promontogno, and then on the right between the latter and Vicosoprano. The Maira is far from having reached its equilibrium curve ; its banks, being undermined, are gradually slipping, thus proving the recent age of this stream.

From Staub's geological map (Bregaglia No. 90) it will be noticed that the Upper Engadine drift covers various landslips. This is a proof that the capture of the Inn by the Maira is older than the last glacial epoch. Furthermore, erratic blocks of Julier granite and of Bernina diorites show that the ice which originated in the Bernina region did not flow entirely towards the Lower Engadine. The low gradient of the valley, already mentioned, compelled the glacier to reascend the valley where the Upper Engadine lakes are now, and then to flow into the Bregaglia. It is a fine example of diffluence. Albert Heim noted long ago that the glacial striæ pointed to the existence of ice motion towards the Maloggia.

CHAPTER X

THE EFFECT OF THE STRUCTURAL SURFACE OF THE PENNINE ALPS ON THE DIRECTION OF THE VALLEYS

CONSEQUENT AND SUBSEQUENT VALLEYS, ON THE SOUTHERN SIDE OF THE SIMPLON-TICINO CULMINATION

Streams and valleys, the courses of which are determined by the original slope of the land (the structural surface), are said to be consequent. Subsequent courses represent the tributary streams of the former. The direction of the consequent courses is thus perpendicular to the isohypses of the structural surface, while the subsequent courses follow them. We have seen that the Simplon-Ticino region represents a culmination of pitch. To the north-east, the axis of the folds lowers itself towards the Grisons and to the south-west it goes down towards the Valais.

If we study the direction of some of the streams, we detect consequent and subsequent courses.

In the Ticino, the Verzasca and Maggia valleys are, at present, transversal valleys. They are derived from ancient consequent courses.

To the east of this median group, the valleys show a tendency to trace curves convex towards the east. On the western side the contrary occurs. These courses are thus almost symmetrical, following the structural isohypses pertaining to the culmination of pitch.

These curved courses were determined by ancient subsequent courses, still recognizable at present. Thus, the isohypses of the previous land-surface were roughly lowered into the present surface of the ground, about twenty kilometres deeper.

The Mesolcina valley is a good example of convexity towards

198

the east. The Toce valley, on the other hand, turns its convexity towards the west.

THE VALLEYS OF THE NORTHERN SLOPE OF THE PENNINE ALPS

According to Argand, the structural surface of what is now the northern slope of the Pennine Alps, was formerly gently dipping towards the north ; its shape was wholly a result of the effect of the Hercynian obstacles on the flow of the nappes.

In the central part of the Valais, the flow of the nappes was easy, owing to the transversal depression situated between the Aar Massif and the Mont-Blanc Massif. The structural isohypses were consequently convex towards the north. Farther to the east, on what was to become the catchment basin of both Visp rivers, the flow of the nappes was hampered, and the structural isohypses modelled themselves to a certain extent, on the round buttress offered by the Aar Massif. The result was that they formed a curve concave towards the north.

The structural isohypses had then a double curvature. They were convex in the western part of the outer Pennine Alps, and concave in the eastern part. Owing to this fact, the primary consequent courses diverged towards the north in the western region and converged towards the north in the eastern region.

This disposition now occurs in the main tributary valleys, on the left side of the river Rhone. From the Isérables valley to that of Grimenz, including the valleys of Nendaz, Hérémence, Réchy, Anniviers, Turtmann, the courses diverge. On the other hand, the valleys of Saas and Zermatt converge.

CHAPTER XI

THE AOSTA VALLEY

The Aosta valley, as shown by Argand, is mostly formed by the Monte Rosa and the Dent Blanche Nappes, in conjunction with their great intermediate Mesozoic lying syncline.

The Gran Paradiso Massif belongs to the Monte Rosa Nappe and Argand admits that it is connected with the Monte Rosa Massif underneath the *Schistes lustrés*, which fill up the region separating both groups of mountains. In fact, it is known that there is a pitching towards the south-west from the Monte Rosa. But on reaching the Dora Baltea, we notice a change of pitch to the north-east away from Gran Paradiso.

On the basis of these observations, we arrive at the conclusion that there is a transversal syncline between Monte Rosa and the Gran Paradiso. The Mesozoic strata covering the subterranean connection of Monte Rosa and the Gran Paradiso Massifs occupy this transversal syncline (Fig. 61).

The massifs of the Dent Blanche, Mont Mary, Mont Emilius and Mont Rafré, as well as the outlier of Mont Pillonnet, are remnants of the Dent Blanche Nappe. They all rest on the Mesozoic strata that overlie the Monte Rosa Nappe. We may thus conclude, with Argand, that the Dent Blanche Nappe has been spared by the erosion, thanks to the transversal syncline of the Aosta valley.

KEY TO FIG. 61.

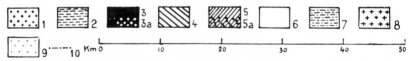

1, Dinarides. 2, Canavese zone. 3, Arolla series of the Dent Blanche Nappe. 3a, Valpelline series of the Dent Blanche Nappe. 4, Monte Rosa Nappe. 5, Crystallines of the Great St. Bernard Nappe. 5a, Carboniferous of the Great St. Bernard Nappe. 6, " Schistes lustrés." 7, High Calcareous Alps. 8, Diorite. 9, Quaternary. 10, Divide.

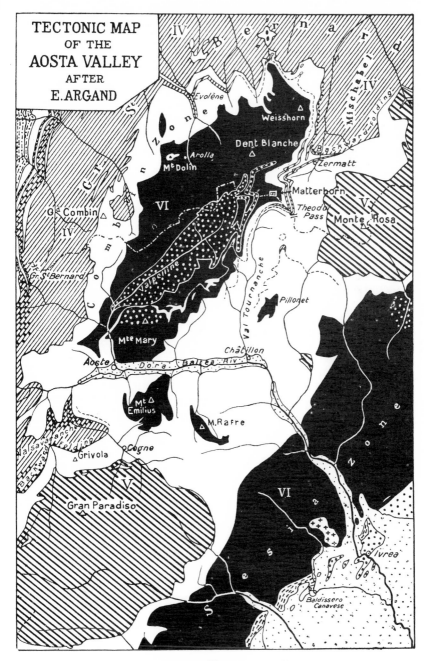

FIG. 61. *For Key see page 200.*

201

The height of the thrust-plane on the left slope of the Aosta valley descends from 3,000 metres to 600 metres, which gives a declivity of 46 in one thousand towards the south-west.

On the southern side of the Dora Baltea, the same phenomenon takes place, *but reversed*. Indeed, both extremities of the Mont Emilius Massif, which is nothing but an outlier of the Dent Blanche Nappe, rest on the *Schistes lustrés*. Here we have an average declivity towards the north-east of 175 in one thousand.

It follows that, on both sides of the valley, from Aosta as far as St. Marcel, over a distance of 10 kilometres, the lower surface of the Dent Blanche Nappe is decidedly pitching towards the river course. The pitch is stronger on the right side than on the left. The depth of the depression thus obtained is at least 2,400 metres. The direction of this tectonic depression being west to east, it is therefore sub-transversal to the direction of folding.

Over a considerable part of its course, the Dora Baltea valley very nearly corresponds with the tectonic depression. We are thus entitled to believe that tectonics have played an important part in determining the valley itself. This genetic interdependence between transversal valleys and transversal synclines has been noticed also by Lugeon in several cases in the High Calcareous Alps.

CHAPTER XII

THE ROOTS OF THE PENNINE NAPPES

General Characters. Between Crevola and Ornavasso, over a distance of 24 kilometres, the river Toce, on Italian territory, crosses the region of the roots of the nappes. From north to south, the following roots occur : those of Antigorio, Lebendun, Monte Leone, Great St. Bernard, Monte Rosa, Dent Blanche ; then come the roots of the Austrides.

In this region the nappes assume a more or less vertical position. Sedimentary zones, consisting of dolomitic marbles and metamorphosed *Schistes lustrés*, separate the nappes. Sometimes these zones (synclines) have been crushed and stretched to such an extent that only lenticles of them appear. When the sedimentary is lacking, thrust-planes mark the contact between two nappes.

The zone of roots of the Pennine Nappes, as is the case for the Austrides, may be followed to the east. It crosses the Swiss territory in the region of Arbedo, to the north of Bellinzona, then trends towards the lower part of the Piano de Chiavenna and the Ratti valley, where it is intercepted by the Tertiary granites of the Bregaglia (see page 195). From there it passes to the south of the Monte della Disgrazia to reach the valley of Poschiavo, to the south of the Bernina Pass, where the junction of the Margna Nappe with its roots occurs.

Dent Blanche Nappe. As shown in Plate XI, Section A, Argand considers the zone of Sesia Lanzo, on the Italian side of the Alps, to be the root of the Dent Blanche Nappe.

Indeed, the outliers of the Dent Blanche Nappe, that is Mont Emilius, Mont Rafré and Mont Pillonnet are remnants of what we may call the bridge uniting the nappe to its roots. The Arolla gneisses and glaucophane rocks of Mont Emilius occur in the zone of Sesia Lanzo. This zone is bordered on

203

the south by the Canavese zone, which separates the Ivrea zone from the Sesia-Lanzo zone. The Canavese zone represents the normal sedimentary cover of the Sesia-Lanzo zone. As will be seen later on, the Ivrea zone corresponds to the roots of R. Staub's Austrides.

Margna Nappe. As shown by R. Staub, the best region to follow the Margna Nappe to its roots is the Passo d'Uer, to the west of the Poschiavo valley (south of the Bernina Pass). Over a distance of 25 kilometres, the crystalline rocks of the root of the Platta digitation of the Margna Nappe are separated from the crystalline rocks of the root of the Margna Nappe (in a restricted sense) by a zone of sedimentary lenticles. These may be followed from Le Prese, on the Poschiavo lake, through the Passo Canciano, Pizzo Scalino, Val di Togno and Monte Foppa, as far as the middle part of the Malenco vale. Farther to the west this zone dies away at the Monte Caldenno and the crystalline rocks unite, forming a single root, which is the eastern equivalent of Argand's Sesia-Lanzo zone.

BIBLIOGRAPHY

1. Schardt, H.—Les résultats scientifiques du percement du tunnel du Simplon. Bull. techn. Suisse romande 1905.
2. Schmidt, C. Preiswerk, H. {Erläuterungen zur geologischen Karte der Simplongruppe. 1 : 50,000. N⁰ 6. Zürich, 1908, 8⁰, 72 pp. fig. pl.
3. Kilian, W.—Etudes géologiques dans les Alpes occidentales. Mém. Carte géol. France, 1905–1908.
4. Argand, E.—La Doire Baltée en aval d'Aoste. Revue de géographie annuelle, T. III. Paris, 1909.
5. Meyer, H.—Geologische Untersuchungen am Nordostrande des Surettamassives im südlichen Graubünden. In. Diss. Ber. d. Naturf. Ges. Freiburg i. B. Bd. xvii, 1909.
6. Argand, E.—L'exploration géologique des Alpes pennines centrales. Bull. N⁰ 14 des Lab. de Géologie, Géog. phys. etc. de l'Université de Lausanne. Lausanne, 1909.
7. Argand, E.—Les nappes de recouvrement des Alpes pennines et leurs prolongements structuraux. Mat. carte géol. de la Suisse. N.S. xxxie Liv. Berne, 1911.
8. Argand, E.—Sur le drainage des Alpes occidentales et les influences tectoniques. Proc.-verb. Soc. vaud. Sc. nat. Lausanne, 3 avril 1912.
9. Argand, E.—Le faîte structural et le faîte topographique des Alpes occidentales. Proc.-verb. Soc. vaud. Sc. nat. Lausanne, 17 avril 1912.

10. ARGAND, E.—Sur la segmentation tectonique des Alpes occidentales. Bull. Soc. vaud. Sc. nat., vol. XLVIII, N° 176. Lausanne, 1912.

11. HAUG, E.—Les nappes de charriage de l'Embrunais et de l'Ubaye et leurs faciès caractéristiques. Bull. Soc. géol. France, 4e sér., T. 12, p. 1, 1912.

12. KILIAN, W. ⎰La série sédimentaire du Briançonnais oriental.
    PUSSENOT, CH.⎱Bull. Soc. géol. France, 4e sér., T. 13, p. 17, 1913.

13. HAMMER, W.—Das Gebiet der Bündnerschiefer im tirolischen Oberinntal. Jahrb. d. K.K. geol. Reichsanst. 1914, Bd. 64, 3. Heft. Wien, 1915.

14. ARGAND, E.—Sur l'arc des Alpes occidentales. Eclogae geol. Helvet., vol. XIV, N° 1, 1916.

15. STAUB, R.—Ueber Faciesverteilung und Orogenese in den sudöstlichen Schweizeralpen. Beitr. z. geol. Karte d. Schweiz. N.F. 46, III, 1917.

16. PREISWERK, H.—Oberes Tessin und Maggiagebiet. Beitr. z. geol. Karte d. Schweiz, 26e Lief. Bern, 1918.

17. KRIGE, J.—Petrographische Untersuchungen im Val Piora und Umgebung. Eclogae geol. Helvet., vol. XIV, H. 5, 1918.

18. PREISWERK, H.—Ueber die Geologie der N.W. Tessineralpen. Verh. d. schw. Naturf. Ges. Lugano, 1919.

19. WILKENS, O.—Beiträge zur Geologie des Rheinwalds und von Vals (Adula-Graubünden). Geolog. Rundschau, Bd. XI, Heft 1/4, 1920.

20. TERMIER, P.⎰Sur l'âge des schistes lustrés des Alpes occidentales.
    KILIAN, W. ⎱C.R. Ac. Sc. Paris, 27 décembre 1920.

21. TERMIER, P.⎰Le lambeau de recouvrement du Mont Jovet en
    KILIAN, W. ⎰Tarentaise ; les schistes lustrés au Nord de Bourg-
             ⎱Saint-Maurice. C.R. Ac. Sc. Paris, 6 déc. 1920.

22. TERMIER, P.⎰Le bord occidental du pays des schistes lustrés dans
    KILIAN, W. ⎰les Alpes franco-italiennes entre la Haute-Maurienne
             ⎰et le Haut-Queyras. C.R. Ac. Sc. Paris, 8 nov.
             ⎱1920.

23. BARTHOLMÈS, F.—Contribution à l'étude des roches éruptives basiques contenues dans le massif de la Dent Blanche. Bull. des Lab. de Geol. etc. de l'Université de Lausanne, N° 27, 93 pp. Genève, 8°, 1920.

24. SANDER, B.—Geologische Studien am Westende der Hohen Tauern. Jahrb. d. geol. Staatsanst. Wien, 1920, pp. 273–296.

25. STAUB, R.—Ueber den Bau des Monte della Disgrazia. Vierteljahrschr. Naturf. Ges. Zürich, 66. 1921.

26. CORNELIUS, H. P.—Ueber einige Probleme der penninischen Zone der Westalpen. Geol. Rundschau 1920/21, T. XI, p. 289.

27. TERMIER, P.—Le bord occidental du pays des schistes lustrés dans la Haute Ubaye. Bull. Soc. géol. France, 4e sér., T. 21, p. 286, 1921 (22).

28. KOBER, L.—Das östliche Tauernfenster. Denkschr. Ak. Wiss. Wien, 1922.

29. WEGMANN, E.—Zur Geologie der St. Bernharddecke im Val d'Hérens (Wallis). (Thèse.) Bull. Soc. Neuchâteloise des Sc. nat. T. XLVII, 1922.

30. TSCHOPP, H.—Die Casannaschiefer des obern Val de Bagnes (Wallis). Eclogae geol. Helvet., vol. XVIII, 1923, p. 77.

31. FRISCHKNECHT, G.—Die zwei Kulminationen Tosa und Tessin und ihr Einfluss auf die Tektonik. Eclogae geol. Helvet., vol. XVII, p. 522, 1923.

32. JENNY, H.
    FRISCHKNECHT, G. ⎰Geologie der Adula. Beitr. z. geol. Karte d.
    KOPP, J. ⎱Schweiz. N.F. 51. Bern, 1923.

33. ARGAND, E.—La géologie des environs de Zermatt. Actes Soc. helvet. des Sc. nat. Zermatt 1923, IIe part. Pp. 96–110.

34. WERENFELS, A.—Geologische Beschreibung der Lepontinischen Alpen. Dritter Teil. Geologische und petrographische Untersuchung des Vispertales. Beitr. z. geol. Karte d. Schweiz. 26e Lief. Bern, 1924.

35. BOSSARD, L.—Geologie des Gebietes zwischen Val Leventina und Val Blenio. In. Diss. Eclogae geol. Helvet., vol. XIX, p. 504, 1925.

36. STAUB, W.—Zur Tektonik des Gebirges zwischen Turtmanntal und Simplonpass. Eclogae geol. Helvet., vol. XX, No 2, 1927.

37. SCHUMACHER, GERHARD.—Geologische Studien im Domleschg (Grabünden). Jahrbuch d. phil. Fak. II, Univ., Bern. Bd. VII, 1927, pp. 157–74.

38. STRASSER, ERNST.—Geologie der Pizzo di Claro-Torrone-Altokette sowie der penninischen Wurzelzone zwischen Val Calanca und Tessintal. Thèse, Zurich, 1928, 71 pp.

39. MÜHLEMANN, RUDOLF.—Geologische und morphologische Untersuchungen im Gebiete der Tambodecke zwischen Val Mesolcina und Valle San Giacomo (Italien). Thèse, Zurich, 1928, 60 pp.

40. GRÜTTER, OTTO.—Petrographische und geologische Untersuchungen in der region von Bosco (Valle Maggia) Tessin. Verhandl. Naturf. Ges. Basel, Bd. XL, 1 teil 1929.

41. CADISCH, J.—2. Tektonik und Stratigraphie im penninisch-ostalpinen Grenzgebiet. Verhandl. Naturf. Ges. Basel, 1929, XL, pp. 62–77.

42. ARBENZ, P.—Crinoidenfunde im Lias der Dent Blanche decke am Mt. Dollin bei Arolla und den Bündnerschirfern der Alp Monterascio sudlich der Greina (Tessin). Eclogae geol. Helvet., vol. XXIII, No 2, 1930.

43. WILHELM, O.—Geologie der Landschaft Schams (Graubünden). Beitr. z. geol. Karte der Schweiz. Nelle sér., 64e Liv., 1932, pp. 1–32.

44. KOPP, J.—Zur Stratigraphie und Tektonik der Gebirge zwischen Lugnezer und Valser Tal (Piz Aulgruppe). Eclogae geol. Helvet., 1933, pp. 191–197.

45. ARGAND, EMILE.—La zone pennique. Guide géologique de la Suisse, 1934. Wepf et Cie, Bâle, p. 149.

46. RÜGER, L.—Zur Altersfrage der Bewegungen und Metamorphosen im Penninikum der Tessiner Alpen. Geol. Runds., Bd. XXV, 1934. H. 1, pp. 1–10.

47. KUNDIG, E.—Neue Gesichtspunkte in den Problemen der Tessiner-Tektonik. Eclogae geol. Helvet., 1934, pp. 333–336.

48. PREISWERK, HEINRICH ⎰Geologische Uebersicht über das Tessin.
    und REINHARD, MAX. ⎱Guide géologique de la Suisse, 1934, p. 190. Wepf et Cie, Bâle.

PART IV

BETWIXT-MOUNTAINS
THE EASTERN ALPS OR THE AUSTRIDES

CHAPTER I

INTRODUCTION

As shown by Pierre Termier, *the Eastern Alps override the Western Alps*.

The Eastern Alps stretch from a line joining the Rhine river to the Septimer Pass, as far as Vienna. From a geological point of view, they override the Western Alps, as shown in Fig. 62. In fact, the *Flysch* of the Pennine Nappes is overlain by sedimentary rocks, as seen at the foot of the great cliff of the Rhätikon, or at Arosa. These rocks represent the lowest tectonic element of the Eastern Alps.

As shown by Staub, the Eastern Alps are made up of a series of tectonic elements, or nappes, which overlap one another.

Termier and Argand used the word " Austro-Alpine Nappes " for the tectonic elements forming the Eastern Alps. R. Staub replaced this expression by the more general word, " Austrides," for these nappes form the Austrian Alps, and he divides them into the following great groups :

The Tirolides or Upper elements.

The Grisonides or Lower elements.

The facies of the Grisonides is a passage-facies between the Pennine Nappes and the true East Alpine facies, represented in the Tirolides. The name of Grisonides is taken from the Canton of Grisons in Switzerland, where the lower elements of the Austrides play a predominant rôle in the scenery.

The Grisonides have been divided by Staub into the following elements, in descending order :

207

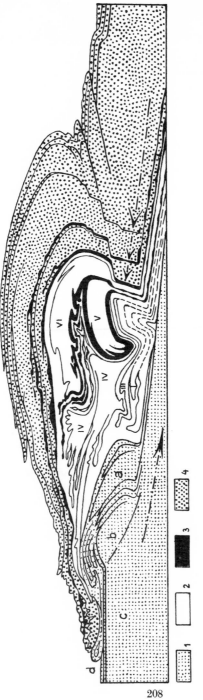

The Eastern Alps override the Western Alps.

Fig. 62.—Generalized Section showing the Structure of the Alps. *After E. Argand.*

1. *Foreland or Eurasia.* a, crystalline wedges; b, swelling of the crystalline; c, crystallines without any deformation.

2. *The Pennine Nappes,* formed into the Alpine geosyncline, which represent the *Western Alps.* I–III, Simplon-Ticino Nappes. IV, Great St. Bernard Nappe. V, Monte Rosa Nappe. VI, Dent Blanche Nappe.

3. Basic rocks.

4. *The Eastern Alps.*

The Campo Nappe.

The Err-Bernina Nappe.

The Tirolides are the nappes out of which the Tirol mountains were carved. They represent the most powerful elements of the Austrides, which were thrust over the Pennine and Grisonides Nappes. It is what Pierre Termier called, twenty-three years ago, the " crushing sledge," or the " traineau écraseur."

In the Tirolides we have to deal with three nappes, forming one huge tectonic element : *The Silvretta Umbrail-Oetztal Nappe.*

At the first glance, the Eastern Alps seem to be of a very complicated nature (see page 4). In fact, the Northern Limestone zone, to which this effect is mostly due, must be considered, as shown by Staub, as a patch of sedimentary slices, all belonging to the Mesozoic cover of a great crystalline element : the Silvretta-Oetztal Nappe.

The tectonics of the Austrides cannot be compared to the tectonics of the Pennine Alps. In fact, we have to deal, in the Austrides, with overthrusts of the type of the Foreland and not with recumbent anticlines of the Pennine type. In reality, the Austrides are " Betwixt-mountains " (see page 25).

That the Austrides have been thrust over the Pennine Nappes and the Foreland, is demonstrated by the existence of two windows :

The window of the Lower Engadine.

The window of the Hohe Tauern.

The Lower Engadine window was eroded out of the Austrides nappes. In fact, its frame is made up of Grisonides and Tirolides and the Pennine *Schistes lustrés* appear in it.

The Hohe Tauern window is built on a much greater scale. The Austrides nappes frame it also, but in the window occur two tectonic elements of the Pennine Alps, viz. the homologue of the Monte Rosa Nappe and of the Dent Blanche Nappe (Margna), surrounded by *Schistes lustrés.*

In a third window, the *Semmering window,* appear the Grisonides framed by the Silvretta Nappe.

As already seen (page 32), the windows occur on culminations of pitch due, without doubt, to massifs of the Foreland which have been overridden by both the Pennine elements and the Austrides.

An understanding of the Eastern Alps is then based on knowledge of the Western Alps.

S.A.

CHAPTER II

THE GRISONIDES

The Grisonides are, according to R. Staub, the lowest part of the Eastern Alps or Austrides. They have been named after the eastern region of the Swiss Alps, the " Canton of Grisons (Graubünden)," where they have been determined and where they form the highest peaks of the Eastern Alps.

The Grisonides represent the junction between the Pennine Nappes and the highest element of the Eastern Alps, the Silvretta Nappe.

R. Staub has divided the Grisonides as follows, in descending order :

The Campo Nappe.

The Err-Bernina Nappe.

The Err-Bernina Nappe rests on the Margna Nappe (homologue of the Dent Blanche Nappe), the highest element of the Pennine Alps.

The Campo Nappe unites with the Err-Bernina Nappe at the Sassalbo, a summit to the east-north-east of the village of Poschiavo, to the south of the Bernina Pass.

The Err-Bernina Nappe

To the Err-Bernina Nappe belong the mighty crystalline massifs of the Upper Engadine, with the Albula granite and the Julier granite.

These crystalline massifs represent the core of the Err-Bernina Nappe, with their digitations *Sgrischus*, *Albula* and *Stretta*.

To the north of the Grisons, the *Falknis* and *Sulzfluh* represent a part of the Err-Bernina Nappe which has been dragged by the travel of the higher nappes of the Eastern Alps. The Falknis and Sulzfluh play a great rôle in Alpine Tectonics, for they constitute the connection between the Prealps, the Klippes and the Eastern Alps.

In the Arosa mountains, as will be seen later on, formations occur which connect the Falknis and Sulzfluh to the Err-Bernina Nappe.

The base of the Grisonides, viz. of the Err-Bernina Nappe, is represented by the Albula granite, which rests on the Platta digitation. of the Margna Nappe.

The name Err-Bernina Nappe expresses the idea of a single tectonic element, but in reality we have to deal with two nappes representing a big tectonic element, as will be seen later on.

THE ERR NAPPE

Rocks of the Core. The crystalline rocks forming the core of the Err Nappe are Casanna schists with grey and green granite. Veins of porphyry traverse the Carboniferous sediments, showing that the granite is of Lower Permian age.

The main part of the nappe, to the north, is represented by the Err Massif made up of the Albula granite ; to the south, by the Corvatsch granite. The junction, between these two parts, from a petrological and tectonical point of view, occurs at the Gravasalvas, a mountain on the northern side of the lake of Sils.

Outcrop. The Err Nappe overrides the Sella digitation of the Margna Nappe forming the Caputschin in the Fex valley. The Piz Corvatsch (3,459 m.) was carved out of this nappe, as well as the Fuorcla Surlej. Only the lower part of the Piz Surlej is made up of the Err Nappe.

To the north of the Inn river, the Piz d'Err (3,395 m.), the Piz Ot and the Bevers vale were carved out of this nappe. The Albula tunnel passes through the green granite of the Albula, and on its northern side, Trias containing gypsum, Liassic breccias and shales have been met with. Lake Palpuogna, due to the solvent action of water on the limestones, occurs in this sedimentary zone.

Tectonics. The sediment-cover of the Err Nappe goes from the Carboniferous and Verrucano (Permian) up to the Cretaceous.

The *Schistes lustrés* (Bündner schiefer) of the Oberhalbstein, of Upper Jurassic age, belong, without any doubt, to the Err Nappe, for, as shown by Staub, *they form synclines in the Piz Corvatsch and Piz Sgrischus*, to the south of St. Moritz. Moreover, they overlie the Albula granite at the Piz Scalotta. The geology of the Piz Corvatsch is thus of paramount importance to separate the *Schistes lustrés* of the Austrides from those of the Pennine Nappes.

The *Albula digitation* of the Err Nappe must be considered as the frontal part of the Sgrischus, which has been detached and carried northwards. The junction between the Albula digitation and the Sgrischus is demonstrated by the presence of the Oberhalbstein *Schistes lustrés*.

The rocks of the Err Nappe are strongly mylonitized. This occurs especially in the crystallines of the Piz Corvatsch, owing to the pressure exerted by the Bernina Nappe.

Sedimentary Rocks. *The stratigraphical sequence* of the sedimentary cover of the Err Nappe consists of Verrucano, Trias, Lias, Middle Jurassic, Upper Jurassic (radiolarites) and Cretaceous.

The Cretaceous of the Err Nappe, called the Saluver series by Cornelius, consists of Neocomian, Gault and "Couches rouges." It is well exposed on the northern ridge of the Piz Nair, above St. Moritz. The conglomerates of the summit, which were considered to be of Upper Cretaceous age by Cornelius, belong, according to Staub, to the Verrucano (Permian) of the Bernina Nappe. Indeed, this Verrucano represents the reversed limb of the Julier digitation of the Bernina Nappe, which overrides the sedimentary cover of the Err Nappe. In fact, a thrust-plane separates the Verrucano slice from the underlying Cretaceous.

However, as yet the puzzle of the sedimentary rocks of the Piz Nair does not seem to have been solved entirely.

Relations between the Err Nappe and the Bernina Nappe. The Err Nappe was separated from its root by the travel of the Bernina Nappe, as demonstrated at the Sella Pass. The Err Nappe is thus isolated between the Margna Nappe and the Bernina Nappe. In Staub's opinion, the Err Nappe was the frontal part of the Bernina Nappe, and the latter has overridden the former.

As shown in Fig. 63, there is no frontal archbend in the Err Nappe, but only frontal slices, demonstrating the crushing effect of the higher nappes on the lower ones. Moreover, the Campo Nappe shows an involution underneath the frontal slices referred to above.

To sum up, *the Err and Bernina Nappes are two big slices which have their origin in the same huge tectonic element.*

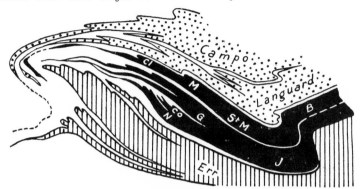

Fig. 63.—The Digitations of the Bernina Nappe in the St. Moritz Region, and the Relations between the Err, Bernina and Campo Nappes. *After R. Staub.*

Bernina Nappe.—B, Bianco Slice.   *St. M*, St. Moritz.   *M*, Musella.   *Cl*, Clavadatsch. *J, Julier Slice.*   *G*, Giop.   *Co*, Corviglia.   *N*, Nair.

THE BERNINA NAPPE

Tectonics. R. Staub has recorded the following slices, which form the frontal part of the Bernina Nappe, above the sedimentary cover of the Err Nappe, in descending order:

Slice of St. Moritz passing into the Bianco slice.

Slice of Giop ⎫
Slice of Corviglia ⎬ Julier slice.
Slice of Nair ⎭

Outcrop. Julier, Lagrev, Munt Arlas, Piz Surlej belong to the Julier slice, which may be followed to the lower part of Piz Tschierva, Morteratsch, Bernina and Roseg.

Piz Bernina, Bianco, Prievlus, Morteratsch, Tschierva, Zupo, Bellavista, Palu, Scersen and Roseg, belong to the Bianco slice.

The Boval corrie is a window of the Julier Nappe, framed by the Bianco slice. The connection between the Boval window and the Julier slice only occurs at the Fuorcla Prievlusa and at the Crastagüzzasattel to the west.

Rocks of the Core. The core of the Bernina Nappe consists of eruptive rocks. Paragneisses, such as Casanna schists, do not play a great rôle ; they only occur on the eastern margin of the nappe.

The principal rocks forming the core of the Bernina Nappe are : diorite, monzonite, banatite (blue granite) and granite (red or white). A very good contact between banatite and Casanna schists may be seen at the Munt Pers, then at the entrance of Val del Fain and at la Pischa. According to R. Staub, the eruptive rocks of the Err Nappe are of an acid type (granite), while the eruptive rocks of the Bernina Nappe are mostly basic (diorites and gabbros).

With the Err Nappe representing the frontal part of the great Err-Bernina tectonic element, and the Bernina Nappe the southern part, we arrive at the conclusion that the acid magma went to the frontal part of the nappe and the basic magma remained near the roots. R. Staub expresses the opinion that marbles may have an effect on the basicity of the magma, for they are always connected, in the Bernina Nappe, with diorites and gabbros.

Sedimentary Cover. The best exposure of *the sedimentary envelope* of the Bernina Nappe occurs at the Piz Alv, where it forms a secondary syncline, to the north of the Bernina Pass. The normal limb of this fold is made up of a thin zone of Trias, and its core is filled up by Rhaetic and Lias, supporting Rhaetic and dolomitic limestones of the Trias, which form the reversed limb. The Lias begins with breccias *as in the outer ranges of the Prealps.* The Trias shows also much likeness to the Median Prealps.

The Mesozoic envelope of the Bernina Nappe extends from St. Moritz, through Pontresina, the Languard vale, la Pischa, Val del Fain, Piz Alv, as far as the Fuorcla Carale, to the south of the Bernina lakes. From the syncline of the Piz Alv, the sedimentary zone envelops the crystalline schists of the Piz della Stretta, and then runs towards the south as far as the Sassalbo, above Poschiavo (Fig. 64).

Stretta Digitation. The crystalline of the Piz della Stretta, enveloped by sedimentary rocks, forms the higher digitation of the Bernina Nappe, called by Staub the Stretta slice.

The Sassalbo sedimentary separates the crystalline of the Stretta slice of the Bernina Nappe from the crystalline of a higher element of the Austrides, namely, the Campo Nappe.

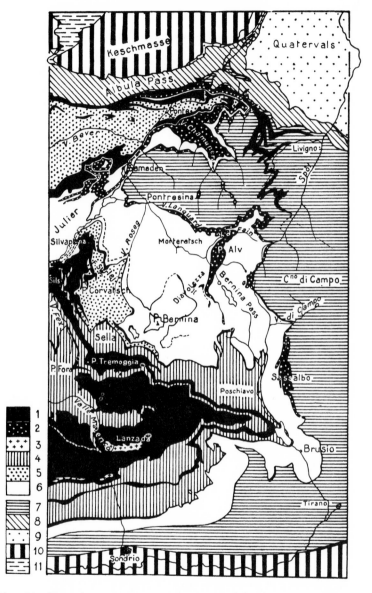

Fig. 64.—Tectonic map of the Region situated between St. Moritz and Poschiavo. *After R. Staub.*

1, Mesozoic in general. 2, Mesozoic of the Bernina Nappe. 3, Suretta Nappe. 4, Margna Nappe. 5, Err Nappe. 6, Bernina Nappe. 7, Campo Nappe. 8, Ortler zone. 9, Engadine Dolomites. 10, Keschmasse. 11, Ducan Trias.

The Err-Bernina Nappe

in the Rhätikon and in the Arosa Region

Tectonics. The Rhätikon commands, to the north, the Prätigau valley. This range consists of the following tectonic elements, in descending order (Fig. 65) :

5. *The Silvretta Nappe*, out of which the Scesaplana (2,969 m.) has been carved.

4. *A thin zone of " Aroser schuppen."*

3. *The Sulzfluh Nappe*, forming the Sulzfluh and Drusenfluh, to the east.

2. *The Falknis Nappe*, out of which the Falknis was eroded, to the west.

1. *The Prätigau Flysch* or *Schistes lustrés* (Bündner schiefer) of Tertiary age, forming the bottom and the lower slopes of the valley.

The *Prätigau Flysch* belongs to the Margna Nappe of the Pennine Alps. It is thus overridden by tectonic elements of the Austrides from the Falknis Nappe up to the Silvretta Nappe. It is in the Rhätikon that the Western Alps disappear underneath the Eastern Alps. The overlap of the Pennine *Schistes lustrés* by the Austrides may be followed to the south, from the Sulzfluh as far as the Arosa region.

The results arrived at by Trümpy, Cadisch, Brauchli and R. Staub show that the following elements must be correlated with the Err-Bernina Nappe, from top to bottom :

Err-Bernina Nappe.
$\left\{ \begin{array}{l} (c)\ \text{The zone of ``Aroser schuppen.''} \\ (b)\ \text{The Sulzfluh Nappe.} \\ (a)\ \text{The Falknis Nappe.} \end{array} \right.$

The following scheme (Fig. 66), after Cadisch, enables us to understand these

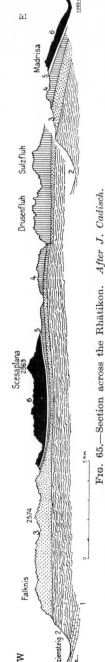

Fig. 65.—Section across the Rhätikon. *After J. Cadisch.*

1, High Calcareous Alps. 2, Prätigau Flysch or " Schistes lustrés." 3, Falknis Nappe. 4, Sulzfluh Nappe. 5, " Aroser schuppen." 6, Silvretta Nappe.

relations. The Falknis and Sulzfluh Nappes are made up of sedimentary rocks of Mesozoic and Tertiary age, showing an orogenic facies. The "Aroser schuppen," on the other hand, consist of paragneisses Verrucano (Permian), Triassic dolomitic limestones, bathyal Lias, limestones with Aptychus, and radiolarites.

This structure in slices (schuppen), which resembles the structure of the External Prealps, is due to the travel of the higher nappes of the Austrides (see Fig. 66).

To sum up, the *sedimentary rocks of the Falknis and Sulzfluh*

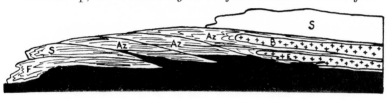

Fig. 66.—Scheme of the Relations between the Falknis-Sulzfluh Nappes and the Err-Bernina Nappes. *After J. Cadisch.*

S, Campo and Silvretta Nappes. *B*, Crystalline core of the Bernina Nappe. *E*, Crystalline core of the Err Nappe. *Az*, "Aroser schuppen." *S*, Sulzfluh Nappe. *F*, Falknis Nappe. In black : the Margna Nappe (Pennine).

Nappes, as well as the sedimentary rocks of the "Aroser schuppen," represent the Mesozoic and Tertiary envelope of the Err-Bernina Nappe. This envelope was detached from its crystalline core and piled up in slices, owing to the northward travel of the higher nappes of the Austrides.

THE ERR-BERNINA NAPPE
IN THE LOWER ENGADINE WINDOW

As shown by R. Staub and Cadisch, the Falknis and Sulzfluh Nappes, as well as the "Aroser schuppen," form the base of the frame of the Lower Engadine window. They rest on the Pennine *Flysch*, which appears in the window, and are overridden by the crystalline of the Silvretta Nappe, as in the Rhätikon.

The results arrived.at by Staub and Cadisch in the study of the Lower Engadine window are of paramount importance. Indeed, these authors found at the south-western corner of the window, at Ardez, *the junction of the sedimentary rocks of the Falknis series with their crystalline substratum, represented by the Tasna granite.*

The Tasna granite was thrust over the Pennine *Schistes lustrés*. It supports Verrucano (Permian), Trias, Lias, Upper Jurassic, then Upper Cretaceous. *The green Tasna granite belongs to the Err Nappe.*

The correlation that had been established elsewhere between the Falknis Nappe and the Err-Bernina Nappe was entirely based on the geometry of the nappes. Here, in the Lower Engadine window, other evidence of it has been found. This record is of great value, as will be seen later on, when dealing with the relations between the Prealps and the Austrides.

If we follow the sedimentary series of Ardez towards the north, that is towards the front of the nappe, we notice the disappearance of the strata, one after the other, from the Lower Trias up to the Upper Jurassic, till the Upper Cretaceous rests directly on the crystalline. This is a splendid demonstration of the geanticlinal structure of the Err Nappe.

From a tectonic point of view, the Ardez sedimentary series is complicated by several repetitions of the strata. In fact, we have to deal with a series of slices, as shown, for instance, at the Piz Minschun. There, three slices of sedimentary rocks with Tasna granite were piled up.

The Ardez series may generally be followed along the southern border of the window from Ardez over Crap Puter, then to the south of Tarasp and Vulpera, as far as Plattamala and Nauders,

The Sulzfluh Nappe, according to Staub and Cadisch, is represented in the highest slices of the Ardez series, such as Valmala, Tschainchels and Chaschlogna.

The frame of the Lower Engadine window, to the north, shows the structure found in the Rhätikon, where the Silvretta Nappe overrides the Falknis and Sulzfluh Nappes. In the southern part of the frame, on the other hand, as far as Nauders, the Engadine Dolomites occur between the Silvretta Nappe and the Falknis-Sulzfluh Nappe.

We shall soon understand the meaning of the Engadine Dolomites, when studying the Campo Nappe.

THE ERR-BERNINA NAPPE

IN THE HOHE TAUERN WINDOW

As shown by R. Staub, the Err-Bernina Nappe occurs at the base of the frame of the Hohe Tauern window, to the west and north. To the west, in the region of the Brenner, the Sterzing quartzites belong to the lowest nappe of the Grisonides.

As shown by Termier, at the Tarntalerköpfe, three slices of Triassic-Jurassic sediments are intercalated between the Silvretta Nappe and the Pennine *Flysch*. The structure of the northern part of the Grisons, i.e. of the Rhätikon and Arosa region, are also met with at the Tarntalerköpfe as far as the Zillertal.

The same feature occurs at the base of the Gasteinerklamm. The equivalent of the Err-Bernina Nappe occurs in the eastern part of the frame of the Hohe Tauern window, at the Radstätter-tauern. This region is of paramount importance, for sedimentary rocks are met with of the facies of the Saluver series, of the Piz Alv and of the Sassalbo. Moreover, as in the Lower Engadine, the transgression of the Cretaceous on the crystalline appears. Thus the characteristics of the lower Grisonides, *i.e.* the Err-Bernina Nappe, may be followed over 300 kilometres from the Upper Engadine as far as the eastern part of the Hohe Tauern window. But the lowest element of the Grisonides extends still farther to the north-east. Let us study :

THE ERR-BERNINA NAPPE
IN THE SEMMERING WINDOW

As shown by Suess, Uhlig, Mohr and Kober, the lowest element of the Grisonides occurs in the Semmering window. Diener and Suess, forty years ago, pointed out that the Trias of the Semmering was very like the Trias of the Bernina Pass.

The coral limestones of the Upper Jurassic are the equivalent of the Sulzfluh rocks of the same age.

THE CAMPO NAPPE

Rocks. In the Campo Nappe eruptive rocks are not so well developed as in the Err-Bernina Nappe. The chief rôle is played by the Casanna schists, which are very .likely of Carboniferous age. The greatest part of the Campo Nappe is thus formed by metamorphosed sedimentary rocks. In the Languard Nappe, biotite-gneiss, chlorite-gneiss, augen-gneiss, phyllites and quartzites with mica are met with. Pegmatite-veins and aplite-veins often traverse these formations. Amongst the intrusive rocks are granite, tonalite, diorite and gabbro. Porphyrite-veins traverse even the Trias in the Ortler region.

Tectonics. The Campo Nappe overrides the Err-Bernina Nappe. It is the greatest tectonic element of the Grisonides. In fact, as shown by Staub, in the Grisons this nappe is limited, to the south, by the line Sondrio-Tonale, to the west by the Poschiavo vale and the Upper Engadine, by the region of Arosa to the north, and the Ortler Massif to the east.

Over the Ortler, the Campo Nappe may be followed as far as the southern border of the Hohe Tauern window. From this region to the east, this nappe is covered by the highest element of the Austrides. It appears again, for the last time, in the Semmering window.

To the south, the Campo Nappe, over a width of 30 kilometres, consists of a single crystalline element which overrides the Mesozoic formations of the Err-Bernina Nappe. However, in the Upper Livigno valley (Italy), the Campo Nappe is divided into the three following digitations, in descending order :

3. The Quatervals Nappe.
2. The Ortler Nappe.
1. The Languard Nappe.

CHAPTER III

THE TIROLIDES

THE SILVRETTA-UMBRAIL-OETZTAL NAPPE

Tectonics. The Tirolides consist of a huge tectonic element : the Silvretta-Umbrail-Oetztal Nappe. This nappe is divided into two digitations, in descending order :

2. The Oetztal Nappe.

1. The Silvretta-Umbrail Nappe.

The Crystalline Rocks. According to Wenk, the rocks of the Silvretta-Umbrail-Oetztal Nappe are characterized by an extraordinary amount of amphibolites. Orthogneisses represent the half of the rocks. Three types of paragneisses are met with : (1) biotite gneisses, (2) gneisses with garnets and staurolite, (3) gneisses passing into hornfels.

Owing to the effect of Alpine folding, the rocks of the lower part of the Silvretta-Umbrail Nappe show a very strong mylonitization.

THE OETZTAL NAPPE

The Oetztal Nappe, like a great carapace, forms the Mur Alps in Styria. The existence of the Hohe Tauern and Semmering windows enables us to understand that we have to deal with a huge thrust mass. Indeed, the Grisonides form the eastern part of the frame of the Hohe Tauern window, and this tectonic element reappears in the Semmering window. The Oetztal Alps and the Mur Alps thus do not root *in situ*, they represent a great overlap.

To the east of the Semmering window the Oetztal Nappe pitches underneath the Hungarian plain.

At the south-western corner of the Mur Alps the Oetztal Nappe unites with its roots, which we will follow, later on, to the west.

The Oetztal Alps, between the Brenner and the Ortler,

have been carved out of the nappe of the same name. Thus, the Oetztal Nappe frames the eastern and south-eastern part of the Lower Engadine window.

THE SILVRETTA-UMBRAIL NAPPE

This nappe has received a double name owing to the fact, demonstrated lately by Wenk on the basis of Hammer's work, that the element is formed of two parts : as shown in Fig. 67. Indeed, the Umbrail mass is the southern part of the Silvretta mass. The separation of the Umbrail mass from the Silvretta mass is due to a phenomenon of underthrusting. Thus the sedimentary cover of the Umbrail crystallines has been peeled off by the travelling of the Oetztal Nappe and accumulated in the depression formed between the two masses. The *Lower Engadine Dolomites* have been carved out of these sedimentaries.

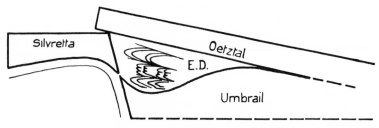

Fig. 67.—The different parts of the Silvretta-Umbrail-Oetztal Nappe.
After E. Wenk.
E.D. Engadine Dolomites.

The Silvretta mass was named after the Silvretta mountains, to the north of the Fluela Pass. This extends farther to the north than the Oetztal Nappe.

On the tectonic map, at the end of the book, we notice that, generally speaking, the Silvretta mass forms the inner border of the Northern Limestone zone.

OUTLIERS RESTING ON THE OETZTAL NAPPE AND THEIR MEANING

The *Tribulaun,* in the Brenner region, is made up of Grisonides and Tirolides. On the Oetztal crystalline, Trias occurs. This formation, as shown by Staub, must be considered as belonging to the sedimentary cover of the Oetztal Nappe. This fact may be followed from Innsbruck as far as the Pflerstal.

Rhaetic and Jurassic may complete the sedimentary series in several cases.

On the sedimentary cover of the Oetztal Nappe there appears, at the Tribulaun, a new tectonic element consisting of Carboniferous :

The Nösslacher Nappe.

The Tribulaun sedimentary is marmorized owing to the pressure exerted by the Nösslacher Nappe. It is a splendid example of dynamometamorphism.

In the Mur Alps, at the Stangalpe, to the west of Turrach, the structure recorded at the upper part of the Tribulaun is also met with. In fact, the Trias of the Oetztal Nappe is over-ridden by Carboniferous.

If there were any doubt whether the Mur Alps belong to the Oetztal Nappe, the presence at the Stangalpe of the Tribulaun structure would solve the question.

On the basis of Tornquist's work, Staub recognizes farther to the east, at Murau, the structure of the Tribulaun and Stangalpe. Moreover, in his opinion, the Devonian of Graz belong to a thrust mass, the Nösslacher Nappe.

To sum up, the Paleozoic rocks of the Graz, Murau, Stangalpe outliers must be correlated with the Nösslacher Nappe of the Brenner. This nappe, the greatest part of which has been eroded away, *thus represents a tectonic element overlapping the Austrides.*

In Staub's opinion, we have to deal with a nappe belonging to the Dinarides, to which he has given the name of :

Styria Nappe.

THE ROOTS

The Roots of the Grisonides. As shown by Staub, the best region to follow the nappes of the Grisonides to their roots is the Passo d'Uer, to the west of the Poschiavo vale. In fact, it is possible to see the flat northern part of the Grisonides uniting with the steep and even overturned roots. Thus the fan-structure of the roots is obvious.

The Err-Bernina Nappe roots in the zone of Brusio. To the east of Ardenno the roots of the Err-Bernina Nappe unite with the roots of the Campo Nappe. Thus we have to deal with only one great tectonic element : *the Grisonides rooting in the Tonale zone.*

The *Tonale zone* may be followed, towards the west, to the *Bellinzona zone* and farther to the *Ivrea zone.*

The most interesting region to study the petrology of the roots of the Grisonides is situated to the near west of Locarno and at Bellinzona itself. In fact, the junction of the Bellinzona zone with the Ivrea zone takes place at Ascona, near Locarno.

The marbles of Ascona separate the diorites of the Monte Verita from the kinzigite [1] schists of Ronco-Brissago. At Bellinzona, the marbles of the Schwyz Castle are met with between a northern band, characterized by the predominance of amphibolites, and a southern band of schists without amphibolites. The Bellinzona zone thus corresponds exactly to the Ivrea zone, and we arrive at the conclusion that *the Tonale zone is the continuation to the east of the Ivrea zone.*

The northern band of the Ivrea zone with diorites resembles the Bernina diorites and the southern band the kinzigite rocks of the roots of the Campo Nappe.

Near Ivrea (Montalto), the *sedimentary zone of the Canavese* occurs as a crushed syncline in the Ivrea zone. This sedimentary separates the roots of the lower Grisonides from the upper Grisonides. In fact, red granite occurs, together with diorites, to the north of the sedimentary zone. These crystalline rocks are characteristic of the Bernina Nappe. Thus the sedimentary of the Canavese represents the sedimentary of the Sassalbo, which separates the Bernina Nappe from the Campo Nappe.

The Tonale zone, to the east, runs through the Ultental mountains over Meran as far as the Mauls gneisses. The roots of the Grisonides are separated from the roots of the Tirolides by a zone of Trias lenticles, running from the Passo San Jorio-Dubino-Ardenno-Monte Padrio-St. Pankraz-Penserjoch and Mauls as far as Kalkstein.

The Roots of the Tirolides. The *Catena Orobica* represents the roots of the Silvretta-Oetztal Nappe. It may be followed, to the west, as far as Monte Cenere, Monte Tamaro, and unites with the *Strona gneisses.* Its southern margin, consisting of a syncline of Carboniferous, Verrucano and Trias,

---

[1] Schists with garnet, sillimanite, andalusite, cordierite and graphite.

separates the roots of the Silvretta-Oetztal Nappe from the Dinarides foundation.

The continuation, to the east, of the Catena Orobica, as roots of the Tirolides, is represented by the Iffinger, the Brixner granite, the Pustertal phyllites, the northern border of the Drauzug as far as the Bacher. In the western part of the Silvretta-Oetztal Nappe the junction with the roots has been removed by erosion. In the eastern part of the nappe, on the other hand, the connection is obvious, as shown in the Mur Alps.

THE NORTHERN CALCAREOUS ALPS

Introduction. The Northern Calcareous Alps, carved out of Kober's Northern Limestone zone of the Eastern Alps, extend from the Rhine river to Vienna. They represent the frontal part of a great tectonic element, the Silvretta-Oetztal Nappe or Tirolides.

The Northern Calcareous Alps override the Grisonides and are even, in some regions, thrust over the Molasse of the Foreland.

Several secondary nappes have been recorded in the Northern Calcareous Alps. Thus, to the west, the Allgäu, Lechtal, Wetterstein, Inntal Nappes. To the east, the Frankenfelser, Lunzer, Oetscher Nappes. On these lower elements rest the nappes of the Salzkammergut and the Juvavian Mass.

Tectonics. R. Staub correlated the results arrived at by the Austrian, French and German geologists, and divided the Northern Calcareous Alps, from a tectonic point of view, as follows, in descending order :

5. The Dachstein Nappe⎫ Juvavian Nappes.
4. The Hallstatt Nappe ⎭
3. The Wetterstein Nappe
2. The Lechtal Nappe⎫ Bavarian nappes of Haug.
1. The Allgäu Nappe ⎭

When dealing with the mode of formation of the Silvretta and Umbrail Nappes (page 222), we have seen that the latter, owing to underthrusting has been separated from its frontal part, the Silvretta Nappe. Study of the relations between the Allgäu, Lechtal and Wetterstein Nappes shows that the Allgäu and Lechtal Nappes represent the sedimentary cover of the Silvretta Nappe, which was peeled off by the travel of the

S.A.

crystalline core of the overlying element, the Oetztal Nappe, and piled up to the north. The Wetterstein Nappe, on the other hand, belongs to the sedimentary cover of the Oetztal Nappe.

After study of the facies of the Hallstatt and Dachstein Nappes, Staub arrived at the conclusion that these nappes must be considered as the sedimentary cover of the Mur Alps (Oetztal Nappe), to the north of the line Stangalpe-Graz.

The Drauzug. The *Drauzug* is a sedimentary zone situated between the roots of the Oetztal Nappe and the Carnian range. The *Drauzug*, according to Staub, does not represent the roots of the Bavarian Nappes, as supposed by several authors. In fact, the deep-sea facies of the Juvavian Nappes must be intercalated between the Bavarian facies and the *Drauzug*.

Staub has summarized the relations between the different tectonic elements of the Northern Calcareous Alps, as follows :

The Allgäu facies of the Silvretta Nappe unites with the facies of the Engadine Dolomites of the Campo Nappe. To the south, the Allgäu facies passes into the Lechtal-Wetterstein facies, which belong to the sedimentary cover of both Silvretta and Oetztal Nappes. The Hallstatt deep-sea facies represents its continuation to the south, where it belongs to the sedimentary cover of the Mur Alps (Oetztal Nappe). It is bordered to the south by the Dachstein facies which represent a passage facies to the Bavarian facies (Allgäu-Lechtal) of the *Drauzug* (see Fig. 68).

To sum up, the *Drauzug* represents the roots of the tectonic elements of the Northern Calcareous Alps, considered as a whole (Plate XI, c).

The Styria Nappe. The *Carnian range*, to the south of the *Drauzug*, made up of Paleozoic rocks, *belongs to the Dinarides foundation*. This Paleozoic represents the roots of the *Styria Nappe*, of which we have found outliers at the Tribulaun, Stangalpe, Murau and Graz. The Trias underlying the Paleozoic of the outliers thus represents the Trias of the Drauzug. We arrive at the conclusion that the *Styria Nappe*, of which only remnants (outliers) have been spared by erosion, *belongs to the Dinarides*.

To sum up, *the Dinarides have overridden the Alps*, as foreseen by Pierre Termier more than twenty years ago.

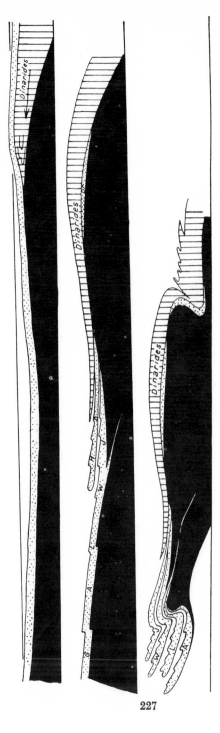

Fig. 68.—Scheme of the Formation of the Northern Calcareous Alps. *After R. Staub.*

Middle Section :—*G*, *Grisonides*. *A*, Sedimentary cover of the *Silvretta* Nappe. *L, W, J*, Sedimentary cover of both *Silvretta* and *Oetztal* Nappes. *H, D*, Sedimentary cover of the *Oetztal* Nappe (Mur Alps). *Dr.* "*Drauzug.*"

Lower Section.—*A*, Allgäu Nappe. *L*, Lechtal Nappe. *W*, Wetterstein Nappe. *J*, Hallstatt and Dachstein Nappes.

Pre-Gosau Orogenic Movements. The facts referred to above enable us to understand the formation of the Juvavian Nappes of the Northern Calcareous Alps. The sedimentary cover of the Mur Alps (Oetztal Nappe) was peeled off by the advance of the Styria Nappe (Dinarides) and was thrust over the lower elements of the Northern Calcareous Alps.

As the *Gosau formation* [1] covers the contact of the Juvavian Nappes, we must admit that the Juvavian push, that is, the overthrust of the Dinarides, is of pre-Gosau age. This phenomenon must be considered as a small orogenic paroxysm, which took place before the great Middle Oligocene paroxysm, to which is due the formation of the whole Alpine range. Argand pointed out, in 1920, that the Western Alps were a mountain range of one paroxysm (Middle Oligocene), and that in the Eastern Alps two orogenic paroxysms made themselves felt, the pre-Gosau paroxysm and the Middle Oligocene paroxysm.

To sum up, the advance of the Styria Nappe is of pre-Gosau age, but the final formation of the underlying nappes of the Austrides, as well as of the Pennine Alps, took place in the Middle Oligocene.

The formation of the Juvavian Nappes of the Northern Calcareous Alps is due to the advance of the Styria Nappe.

CONCLUSIONS

The Importance of the Hohe Tauern Window for Alpine Tectonics

Long ago, the region situated between the Brenner and the Katschberg attracted geologists, for, over 150 kilometres, high mountains occur (Grossglockner, 3,798 m.). In 1854, Sturr and Peters considered the gneisses of the central part of this region as rooting *in situ*, they called them the "*central gneisses.*" The schists which surround them, the "*schieferhülle,*" were supposed to be of Paleozoic age.

At the same time, Studer compared the "*schieferhülle*" with the "*schistes lustrés*" (Bündner schiefer) of the Swiss Alps. In 1890 Suess expressed the opinion that the "*schieferhülle*" was of Mesozoic age.

[1] The Gosau formation represents the transgression of the sea, of Turonian age, after an emergence due to orogenic movements.

Becke, Grubenmann and Weinschenk studied the "*central gneisses*" from a petrological point of view. Weinschenk even considered the granite of this region as a Tertiary intrusion into the "*schieferhülle.*" This mistake was due to the fact that the "central gneisses" were compared to a granitic batholith.

Pierre Termier, in 1903–1906, *pointed out that the* "**central gneisses**" *with the* "**schieferhülle**" *were a window in the Eastern Alps.* He explained, in a general way, the structure of the region. Suess and Uhlig followed suit. Under Uhlig's supervision, detailed mapping was undertaken, and Kober, as Uhlig's successor, took up the work. This author arrived at the result that Termier's hypothesis of the Tauern window was right (Plate XI, c).

In 1920, Kober recorded Pennine nappes in the window. In his opinion, the core of the nappes is formed by the "*central gneisses*" and the "*schieferhülle*" represent their Mesozoic envelope, or the "*schistes lustrés*" of the Pennine Alps.

The confirmation, by Kober, of Termier's hypothesis was of paramount importance for Alpine tectonics. In fact, it was the beginning of the understanding of the geological relations between the Western and the Eastern Alps.

But the splendid work done by the Austrian school of Geology had to be correlated with the results arrived at by the Swiss school. This could only be done by starting from the western part of the Eastern Alps, owing to the pitching towards the north-east. This was R. Staub's standard work, clearly summarized in his *Tektonische Karte der Alpen* (1923).

In the foregoing chapters we have seen that the upper tectonic elements of the Pennine Nappes, viz. the Monte Rosa and Dent Blanche Nappes, must exist underneath the Austrides, for they reappear in the Hohe Tauern window. Moreover, the Grisonides, which form the lower tectonic elements of the Austrides, appear at the base of the frame of the Hohe Tauern window and occur in the Semmering window. *The Pennine nappes, as well as the Grisonides, are overridden by the Tirolides.*

If, at present, we understand the relations between the Western Alps and the Eastern Alps, it is due to Termier's great working hypothesis of the Tauern window utilized and developed by both the Austrian and Swiss schools of geology.

BIBLIOGRAPHY

1. AMPFERER, O. {Geologisches Querschnitt durch die Ostalpen vom
   HAMMER, W. {Allgäu zum Gardasee. Jahrb. d. K.K. geol.
   {Reichsanst., Wien, 1911, Bd. 61. 3. u. 4. Heft.
2. KOBER, L.—Der Deckenbau der östlichen Nordalpen. Denkschr.
   Kaiserl. Ak. Wiss. Wien, 1912.
3. KOBER, L.—Ueber Bau und Entstehung der Ostalpen. Mitt. geol.
   Ges. Wien, 1912.
4. DAL PIAZ, G.—Studii geotettonici sulle Alpi orientali. Mem. Ist
   geol. Univ. Padova, 1912.
5. KOBER, L.—Alpen und Dinariden. Geol. Rundschau Leipzig u.
   Berlin, 1914.
6. SCHWINNER, R.—Analogien im Bau der Ostalpen. Centralbl. f.
   Min. Geol. Pal. Stuttgart, 1915.
7. HERITSCH, F.—A. Die österreichischen u. deutschen Alpen bis zur
   Alpino-Dinarischen Grenze (Ostalpen). Handb. d. reg. geol.,
   II. Bd. 5. Abt. H. 18, 153 pp., fig. pl., 8º, Heidelberg, 1915.
8. STAUB, R.—Geologische Beobachtungen am Bergellermassiv.
   Viertelj. Schr. d. Naturf. Ges. Zürich. Jahrg. 63 (1918). Zürich,
   8º, 18 pp., fig.
9. TERMIER, P.—Sur la structure des Alpes orientales. C.R. Ac.
   Sc. Paris, T. 175, pp. 924, 1173, 1366, 20 nov., 11 et 28 déc.
   1922.
10. KOBER, L.—Das östliche Tauernfenster. Denkschr. Ak. Wiss.
    Wien, 1922.
11. HERITSCH, F.—Geologie von Steiermark. Naturwiss. Ver. f.
    Steiermark. Graz, 1922.
12. SCHWINNER, R.—Die Niederen Tauern. Geol. Rundschau, 1923.
13. KOBER, L.—Bau und Entstehung der Alpen. Borntraeger, Ber-
    lin, 1923.
14. KOSSMAT, F.—Bemerkungen zur Entwicklung des Dinaridenpro-
    blems. Geol. Rundschau, xv, p. 145, 1924.
15. TERMIER, P.—Les nappes des Alpes orientales et la synthèse des
    Alpes. Bull. Soc. géol. Fr., 4 sér., vol. III, 1903, pp. 711–765.
16. AMPFERER, O., und {Geologischer Querschnitt durch die Ostalpen
    HAMMER, W. {vom Allgäu zum Gardasee. Jahrb. geol. Reich-
    {sanst., 1911, Bd. 61, Ht. 3 et 4, pp. 531–710.
17. SANDER, B.—Zur Geologie der Zentralalpen. I. Alpino-dinarische
    Grenze in Tirol ; II. Ostalpin und Lepontin ; III. Stand der
    Deckentheorie in den Alpen. Verh. geol. Reichsanst., Wien,
    1916, Nos 9 et 10.
18. SANDER, B.—Zur Geologie der Zentralalpen. Mit Beiträgen von
    Ampferer O. u. Splenger E. Jahrb. geol. Reichsanst., Wien,
    LXXI, 1921, pp. 173–224.
19. STAUB, R.—Der Bau der Alpen. Beitrage geol. Karte d. Schweiz.,
    N.F., 52 Lief. Bern, 1924.
20. JENNY, H.—Die Alpine Faltung. Borntraeger, Berlin, 1924.
21. ARNI, P.—Geologische Forschungen im mittleren Rhätikon. Thèse,
    Zurich, 1926, 85 pp.
22. STAHEL, A. H.—Geologische Untersuchungen im N.E. Rhätikon.
    Thèse, Zurich, 1926, 82 pp.

23. EGGENBERGER, H.—Geologie der Albulazone zwischen Albulahospiz und Scanfs (Graubünden). Eclogae geol. Helvet., vol. XIX, N° 3, 1926, p. 523.

24. RÖSLI, F.—Zur Geologie der Murtirölgruppe bei Zuoz (Engadin). Jahrb. d. phil., Fak. II, Univ. Bern, 1927, Bd. VII, pp. 140–156.

25. HEGWEIN, W.—Beitrag zur Geologie des Quatervalsgruppe im Schweiz Nationalpark (Graubünden). Jahrb. d. phil., Fak. II, Univ. Bern, 1927, Bd. VII, pp. 98–112.

26. STRECKEISEN, A.—Geologie und Petrographie der Flüelagruppe. Verhand. Schw. Naturf. Ges. Basel, 1927, II teil, pp. 168–169.

27. CADISCH, J.—Geologische Beobachtungen im Gebirge zwischen Unter-Engadin und Paznaun (Tirol). Eclogae geol. Helvet., vol. XXI, N° 1, 1928, pp. 6–7.

28. RÖSLI, F.—Über das gegenseitige Verhältnis von Languard- und Campodecke. Eclogae geol. Helvet., vol. XXI, N° 1, 1928, pp. 9–12.

29. CORNELIUS, H. P.—Zur Auffassung des westlichen Ostalpenrandes. Eclogae geol. Helvet., vol. XXI, N° 1, 1928, pp. 157–163.

30. HERITSCH, F.—The Nappe Theory in the Alps. Translated by Boswell, P. G. H. 1928, Methuen & Co., London.

31. VERDAM, J.—Geologische Forschungen im nördlichen Rhätikon. Thèse, Zurich, 1928, 86 pp.

32. LEUTENEGGER, W. O.—Geologische Untersuchungen im mittleren nordöstlichen Rhätikon. Thèse, Zurich, 1928.

33. KLEBELSBERG, R.—Geologischer Führer durch die Südtiroler Dolomiten. Sammlg. geol. Führer herausgegeben v. E. Krenkel, Berlin. Borntraeger, 1928.

34. CORNELIUS, H., und FURLANI CORNELIUS, M. ⎰Die insubrische Linie von Tessin bis zum ⎱Tonalepass. Denkschr. Ak. Wiss., Wien, ⎰102 Bd., 1930, pp. 207–301.

35. ANGEL, F., und HERITSCH, F. ⎰Das Alter der Zentralgneiss der Hohen Tauern ⎱Centralbl. f. Min. geol., 1931, Abt. B, N° 10, pp. ⎰516–527.

36. WILHELM, O.—Geologie der Landschaft Schams (Graubünden). Mat. Carte géol. Suisse, Nouv. sér., Liv. 64, 1932.

37. BEARTH, P.—Geologie und Petrographie der Keschgruppe. Schweiz. Min. u. Petr. Mitt., Bd. XII, 1932, pp. 256–279.

38. SPAENHAUER, F.—Petrographie und geologie der Grialetsch-Vadret-Sursura Gruppe. Schweiz. Min. u. Petr. Mitt., Bd. XII, 1932, pp. 27–146.

39. KLEBELSBERG, R.—Grundzüge der Geologie Tirols. München, 1933.

40. WENK, E.—Beiträge zur Petrographie und Geologie des Silvretta-kristallins. Schweiz. Min. u. Petr. Mitt., Bd. XIV, 1934, pp. 195–278.

41. WENK, E.—Der Gneiszug Pra Puter-Nauders im Unterengadin und das Verhältnis der Umbraildecke zur Silvretta-Oetztaldecke. Eclogae geol. Helvet., 1934, pp. 134–146, vol. XXVII.

42. DAL PIAZ, G.—Studi Geologici sull'alto Adige orientale e regioni limitrofe. Memorie Inst. Geol. Padova, vol. X, 1934.

43. STAUB, R.—Uebersicht über die Geologie Graubündens. Guide géol. de la Suisse, 1934, Wepf et Cie, Bâle, p. 205.

PART V

THE PROBLEM OF THE PREALPS

CHAPTER I

THE PREALPS

INTRODUCTION

The Prealps are a great patch of mountains, protruding in front of the Alps from the Lake of Thun to the River Arve. From a geographical point of view the Prealps are divided into two parts :

1. The *Romande Prealps*, extending from the Lake of Thun to the Lake of Geneva.

2. The *Prealps of the Chablais*, from the Lake of Geneva to the River Arve.

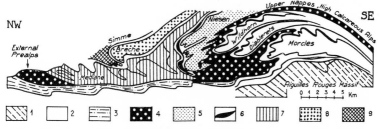

FIG. 69.—Section showing the Nappes of the Prealps.   *After E. Gagnebin.*

1, Massif of the Aiguilles Rouges.   2, Lower Nappes of the High Calcareous Alps.   3, Tertiary of the Swiss Plateau (Molasse).   4, Upper Nappes of the High Calcareous Alps, forming the External and Internal Prealps.   5, Nappe of the Niesen.   6, Sub-Median Belt.   7, The Nappe of the Median Prealps.   8, The Brèche Nappe.   9, The Simme Nappe.

From a geological point of view, the Prealps are a travelled mass entirely foreign to the district in which they stand.   On their southern border they are lying upon the frontal folds of the High Calcareous Alps, and at their northern margin they overlie the *Molasse*, or the Tertiary sandstones of the Swiss Plateau.

Though there are Mesozoic and Tertiary rocks in both the High Calcareous Alps and the Prealps, as shown by fossils their *facies* are quite different. The only possible explanation is that this overlying mass was formed through sedimentation in another country. We shall see, later on, that the higher Prealps, that.can be seen from Geneva, Lausanne and Berne, represent Lower Austrides and a Pennine Nappe.

The Prealps are made of different nappes piled up on one another. The facies, though concerning rocks of the same age, are different in each nappe.

If we cross the Prealps from the Swiss Plateau to the High Calcareous Alps, we find the following tectonic elements in the **Romande Prealps :**

1. *The External Prealps*, at the northern margin, overlying the Swiss Plain.

2. *The Median Prealps*, forming the skeleton of the Prealps. They rest on the former.

3. *The Zone of the Simme*, appearing in Tertiary synclines of the outer ranges of the Median Prealps.

4. *The Zone of the Brèche*, overlying the Median Prealps in their inner ranges.

5. *The Zone of the Niesen*, overlain by the southern margin of the Median Prealps.

6. *The Internal Prealps*, or the Zone of Passes, resting upon the northern border of the High Calcareous Alps and overlain by the Zone of the Niesen. The Internal Prealps represent the southern border of the Prealps. They have also been called the Zone of Passes (Zone des Cols), for all the passes leading from the Lake of Thun to the Lake of Geneva, along the northern border of the High Calcareous Alps, like the Hahnen-moos (1,937 m.), the Trütlisberg (2,040 m.), the Krinnen (1,660 m.), the Pillon (1,546 m.), the Col de la Croix (1,735 m.), have been eroded out of the soft Triassic rocks of this zone.

Plate XI, Section A, shows the tectonic relations between these zones. The External and Internal Prealps belong to the same nappe, the lowest of the Prealps. This nappe has been laminated owing to the travelling northwards of the Median Prealps. The frontal part has been taken away and pushed in front of the Median Prealps where it forms the External Prealps, as shown by Schardt in 1894.

To sum up : We recognize five great tectonic elements in the Prealps which are, from the bottom to the top :

1. The External and Internal Prealps, belonging to the uppermost part of the High Calcareous Alps (see page 92), as shown by Lugeon in 1901.

2. The Zone of the Niesen Flysch.

3. The Median Prealps.

4. The nappe of the Brèche.

5. The nappe of the Simme.

In the **Prealps of the Chablais** we find the same elements, but we notice differences in the details. For instance, on the southern border the nappe of the Brèche is only separated from the High Calcareous by *slices* belonging to the Median Prealps. Thus we arrive at the conclusion that on the southern border of the Prealps of the Chablais, the nappe of the Median Prealps and the Internal Prealps [1] have been almost entirely laminated by the forward movement of the nappe of the Brèche and the nappe of the Simme. These relations are shown in Fig. 70.

[1] Collet and Lillie have just discovered (1935) that the Internal Prealps exist.

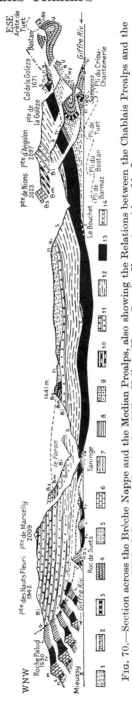

FIG. 70.—Section across the Brèche Nappe and the Median Prealps, also showing the Relations between the Chablais Prealps and the High Calcareous Alps in the Giffre Valley (Haute Savoie, France). *After M. Lugeon.*

High Calcareous Alps (to the right of the section),—1, Flysch (Fl).   2, Nummulitic (N).   3, Upper Cretaceous (C).   4, Barremian, Urgonian facies (U). 5, Neocomian (Ne).

Median Prealps (to the left of the section).—1, Flysch (Fl).   3, Upper Cretaceous (C).   6, Upper Jurassic (M).   7, Dogger (D).   8, Lias (L).   13, Trias   (Tr).

Brèche Nappe (in the middle of the section).—1, Flysch (Fl).   9, Upper Breccia (Bs).   10, Slates (Bm).   11, Lower Breccia (Bi).   12, Shales (S).   13, Trias (Tr).   14, Carboniferous (Ca).

THE EXTERNAL PREALPS

Introduction. The first Alpine range which rises on the south-east border of the Swiss Plateau, from the River Arve to the Lake of Thun, belong to the External Prealps. The principal summits of this range are, from the River Arve to the Lake of Geneva, the Collines du Faucigny and the Voirons near Geneva. From the Lake of Geneva to the Lake of Thun we find : the Pléiades, above Vevey, the Corbettes, the Niremont, the Montsalvens, the Berra and the Gurnigel.

Tectonics. With the exception of Montsalvens, where the tectonics are relatively simple, the External Prealps represent a very complicated zone. Their rocks occur in slices, often intensively crushed, separated by thrust planes.

Stratigraphy. The vegetation covers a good deal of the ground, and renders stratigraphical studies very difficult. It is only lately that Gagnebin succeeded in establishing the stratigraphical sequence of this zone, as follows :

1. Bajocian. Sandy limestones and shales with *Stepheoceras humphriesi.*

2. Bathonian. " Klaus formation " with *Lytoceras tripartitum* and *Perisphinctes procerus.*

3. Callovian. ⎫ Black marly shales containing ferruginous nodules.
4. Oxfordian. ⎭ *Hecticoceras lunula, Cardioceras cordatum.*
All these strata are well exposed in the region of Bulle, where they were discovered by Gilliéron.

5. Argovian. Cement limestones at the base and rugged shales at the summit. *Peltoceras transversarium.* 30 metres.

6. Sequanian. Compact limestones with *Perisphinctes achilles* at the base and rugged limestones with *Peltoceras bicristatum* at the upper part. 10 metres.

7. Kimeridgian. Good bedded limestones with *Aspidoceras acanthium.* 150 metres.

8. Portlandian. Light grey limestone sometimes with conglomerates at its upper part.

9. Berriasian. Often beginning with a basal conglomerate. Grey marls with irregular intercalations of limestones. 40–60 metres.

10. Valanginian. Regular alternations of marly limestones and marly shales. 60–80 metres.

11. Hauterivian. At the base bedded siliceous limestones succeeded by alternations of limestones and shales with *Crioceras duvali.* The latter formation cannot be separated from the :

12a. Lower Barremian. The plasticity of the lower strata causes duplications and secondary folding rendering a detailed stratigraphy impossible, though ammonites occur in several beds.

12b. Upper Barremian (Urgonian facies). Oolitic limestones with *Orbitolina conulus.* 30 metres.

13. Aptian ? Calcareous bedded greensand. The Aptian age of these strata is not certain, the fossils found being undeterminable. 15 metres.

14. Gault ? Dark blue-grey marls.

15. Turonian. Green marls, chalky limestones with Globigerina, green and red limestones with Foraminifera. At the Montsalvens these strata seem to overlie stratigraphically the marls of the Gault.

16. Senonian. Marly limestones with *Rosalina linnei.* The relations of these strata to the Turonian are not visible.

17. Maestrichtian. " Wang formation " with a Serpula, *Jereminella pfenderae. These strata are characteristic of the Wildhorn Nappe of the High Calcareous Alps.* Lugeon has shown the importance of the discovery made by Gagnebin of these strata in the External Prealps —for they occur also in the Internal Prealps—proving the relations established from a tectonic point of view between the External and the Internal Prealps.

18. Nummulitic. This formation consists of : (1) *Black Flysch* at the base, (2) *Wildflysch* and (3) *Sandstones of the Flysch.*

The *Black Flysch*, made up of black micaceous slates, overlies the Molasse (Tertiary) of the Swiss Plateau. It contains Helminthoids and plant remains. Intercalations of limestones with Lithothamnium are also met with. The presence of *Assilina exponens* shows that this formation is of Upper Lutetian or Auversian age. The Black Flysch passes gradually to :

Wildflysch. This " wild Flysch " is made up of black, red and green micaceous shales, lenticles of sandstones, grey and green quartzites, beds of limestones. With Lugeon, we may say that the Wildflysch represents alternations of a terrigenous and pelagic sedimentation.

In the region situated between the Lake of Geneva and the Niremont, slices of Jurassic, Neocomian and Turonian limestones are tectonically intercalated in the Wildflysch, as shown by Gagnebin. Exotic blocks have never been found in it in this part of the External Prealps. As shown by Tercier, the Wildflysch is well exposed to the south-west of the summit of the Berra, a range extending to the north-east of the mountain of Montsalvens, in the Fribourg region. This Wildflysch contains blocks and pebbles called *exotic*, for they have no relation to the rocks forming the mountains on the northern slope of the Alps. These blocks are parts of stacks or debris of shore cliffs, which have fallen into the sea. *They belong to crystalline rocks which formed the shore of the sea* on the bottom of which they were deposited. The petrological character of these blocks will enable us to study their origin.

In the Wildflysch of the north-east part of the External Prealps, we may also find coarse and fine sandstones, called " Gurnigel sandstones."

According to Tercier, the petrology of the exotic pebbles and blocks of the Wildflysch shows that they are derived from the Err-Bernina Nappe of the Eastern Alps (Austrides). The deposition of the Wildflysch was rapid, for several thousand metres of deposits have been formed during the Lutetian and Auversian, as shown by Nummulites. We may try to reconstruct the paleogeography. Since the Wildflysch belongs to the upper nappes of the High Calcareous Alps, the sea in which it was deposited stretched to the south of the High Calcareous Alps, that is along the northern border of the Alpine geosyncline. If the exotic pebbles and blocks of the Wildflysch belong to the Eastern Alps, we must come to the conclusion that the nappes of the Eastern Alps (Austrides) were already thrust over the Pennine nappes. This is a very interesting but bold hypothesis, put forward lately by Tercier. Lugeon, on the other hand, thinks that the exotic pebbles and blocks are derived from a cliff which emerged from the sea, to the south of the region in which the High Calcareous Alps were deposited. This cliff, made up of crystalline rocks, must subsequently have been over-lapped by the travelling Pennine Nappes.

We may hope that detailed studies in the External Prealps will solve the very complicated problem of the formation of the Wildflysch.

Sandstones of the Flysch. On the Jurassic and Neocomian slices referred to above, a very important formation, consisting of inter-calations of shales and sandstones, is met with. At the southern part of the Pléiades *Orthophragmina discus* has been found, and on the southern slope of the Niremont, Gagnebin discovered *Nummulites striatus, Nummulites partschi, Assilina exponens.* These formations belong to the Auversian, and perhaps also to the Priabonian.

THE INTERNAL PREALPS

General Characters. The Internal Prealps represent the most complicated zone of the Alps from a tectonic point of view. Owing to the northward travel of the higher nappes of the Prealps the rocks have been crushed, and it is often almost impossible to identify them. Slices and lenticles separated by thrust planes are very common. As the Internal Prealps are a zone of pastures, outcrops are seldom met with, and it is very difficult to establish any relations between them.

Nevertheless Lugeon succeeded in determining three tectonic elements or three nappes. From the bottom to the top they are :

1. *The Plaine Morte Nappe*, which is the upper digitation of the Wildhorn Nappe of the High Calcareous Alps (see page 93).

2. *The Mont Bonvin Nappe.*

3. *The Oberlaubhorn Nappe.*

When dealing with the High Calcareous Alps (page 92) we arrived at the result that these Internal Prealpine nappes belong to their upper part. They root on the right side of the River Rhone.

Tectonics. As in the External Prealps we often find tectonic intercalations of *Wildflysch* (Tertiary). The Wildflysch belong stratigraphically to the Internal Prealps as well as to the External Prealps, but the stratigraphical relations to the older rocks have not yet been ascertained. The different preliminary papers published on the Internal Prealps by Schardt, Quereau, Lugeon, Roessinger, Sarasin and Collet, Bernet, Arnold Heim do not allow us to solve this puzzle.

Fig. 37 shows the relations between the three nappes of the Internal Prealps and the frontal part of the High Calcareous Alps to the south of Lenk.

R. Staub pointed out in his work *Der Bau der Alpen* that the Trias of the Oberlaubhorn Nappe, with its large proportion of gypsum and its great thickness of dolomitic limestones, cannot belong to the High Calcareous Alps. According to this author the rocks of the Oberlaubhorn Nappe represent the frontal part of the Great St. Bernard Nappe thrust over the High Calcareous Alps by the northward travel of the higher nappes of the Prealps. He calls the Oberlaubhorn Nappe, *the Zone of Bex*. Until a detailed map of the Internal Prealps has been brought out, it will be difficult to express a view on Staub's statement. In the present state of research we may consider Staub's view as a working hypothesis.

THE NIESEN FLYSCH [1]

Introduction. The Niesen (2,367 m.) is the great pyramid which overlooks the Lake of Thun from the south-west. It separates the lower part of the Simme valley from the lower part of the Kander valley. It is the first summit of a range, the Niesen range, running towards the south-west as far as the Albristhorn, to the west of Adelboden. This range borders the Engstligen valley and the Frutigen valley to the north-west. Farther to the south-west, the Niesen range may be followed as far as the Grande Eau valley, where the Pic Chaussy has been carved out of it.

From a stratigraphical point of view, the Niesen range is entirely made up of the *Niesen Flysch*, consisting of breccias from the bottom to the top. In consequence, torrents develop

[1] See Fig. 69.

with great regularity on its slopes, and it is a splendid region to study the formation of a catchment basin.

From a tectonic point of view, the Niesen range rests on the Internal Prealps. This overlap is well exposed above the village of Adelboden and at the Hahnenmoos, the pass between Adelboden and Lenk.

The Niesen Flysch is overridden by the nappe of the Median Prealps, as demonstrated by the overlap of the Spielgerten, to the west of Zweisimmen, in the Simme valley.

Stratigraphy. The Niesen Flysch is a formation 1,000 to 2,000 metres thick. Its principal constituents are polygenous breccias and conglomerates, with intercalations of sandstones, shales and limestones. These strata pass into each other, and though presenting great irregularity along the range, they give an extraordinary monotony to the scenery.

Schardt discovered, in 1898, a specimen of *Inoceramus* at the Arbenhorn. Then Roessinger and Stuart Jenkins, in 1903, ascertained that part of the Niesen Flysch was of Mesozoic age, owing to the presence of a Belemnite in the cement of breccias. Jaccard, in 1906, found also in the cement of breccias of the Chaussy region, a Belemnite, and Lugeon had the same luck near the Lake Liozon.

In 1912, Boussac in his great work on the *Nummulitique Alpin* expressed the opinion—as a working hypothesis—that the Niesen Flysch was a *comprehensive series from the Upper Lias up to the Middle Nummulitic.*

Lugeon showed, in 1914, that Boussac's opinion of a comprehensive series could not be accepted. In fact, he found pebbles of Barremian limestones in the conglomerates of the base of the Niesen Flysch series. The Flysch then could not begin earlier than the Upper Cretaceous. Moreover, in the basal conglomerates of the Niesen Flysch, the crystalline pebbles belong to the Casanna schists of the Great St. Bernard Nappe, while, in the upper part, blocks of granite occur in the sandstones. These facts show that we have to deal with a formation deposited during a transgression which occurred after a phase of emergence.

Tectonics. It was very difficult to express an opinion as to the origin of the Niesen Flysch, from a tectonic point of view, until Lugeon, in 1914, pointed out that this formation was *the frontal part of a Pennine nappe, viz., of the Great St.*

Bernard Nappe, dragged by the travelling of the upper nappes of the Prealps. This opinion is based on a very interesting exposure situated on the road, 1 kilometre to the north of Gsteig, above Gstaad. **Lugeon** found the following succession of strata, in ascending order :

1. Green Casanna schists of the Great St. Bernard Nappe.—5 metres.
2. Crushed quartzites of the Lower Trias.—3 metres.
3. Thin-bedded limestones, slightly dolomitic of the Middle Trias.—6 metres.
4. Dolomitic limestones of the Middle Trias.—2 metres.
5. Green clays of the Upper Trias.—0·5 metres.
6. Grey Liassic limestones.—1 metre.
7. Coarse breccia, made up of pebbles of limestones and crystalline schists, passing into the well-bedded Niesen Flysch.—15 metres.

Argand has lately (1934) shown that Lugeon was right in correlating the Niesen Flysch with the Great St. Bernard Nappe. Indeed, Argand admits that the gneissic slice of the Niesen Flysch belongs to the Middle digitations of the Great St. Bernard Nappe.

Lugeon pointed out that the Niesen Flysch occurs in the Chablais Prealps, at the Chatillon pass, leading from the Giffre valley to the Arve valley. This is the only exposure met with in the Chablais Prealps. Moreover, Lugeon thinks that the Aiguilles d'Arve (France) may be an equivalent to the Niesen Flysch and thus belong to the Great St. Bernard Nappe.

To sum up, we may say that in the present state of research, *the Niesen Flysch must be considered as the frontal part of the Great St. Bernard Nappe dragged by the travelling of the upper nappes of the Prealps.*

THE MEDIAN PREALPS

Introduction.

Northern margin. At many places to the east of the Voirons, it is possible to see the Trias of the Median Prealps resting on the Flysch (Tertiary) of the External Prealps. Between St. Gingolph and Bouveret, on the southern shore of the Lake of Geneva, the Molasse (Tertiary sandstone of the Swiss plain) appears under the Trias.

On the northern shore of the Lake of Geneva, in the neighbourhood of Montreux-Veytaux, the Trias of the Median

s.a.

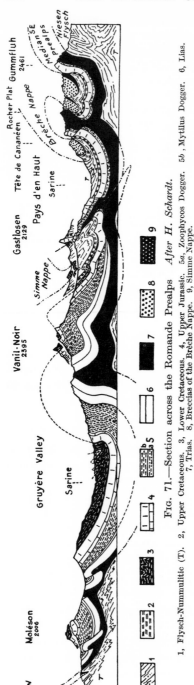

FIG. 71.—Section across the Romande Prealps   *After H. Schardt.*

1, Flysch-Nummulitic (T).   2, Upper Cretaceous.   3, Lower Cretaceous.   4, Upper Jurassic.   5a, Zoophycos Dogger.   5b, Mytilus Dogger.   6, Lias.   7, Trias.   8, Breccias of the Brèche Nappe.   9, Simme Nappe.

Prealps overlies the Flysch (Tertiary) of the External Prealps. The same tectonic contact can be followed farther to the north-east.

Southern margin. From the Kander valley, near Wimmis (Lake of Thun), the Median Prealps rest on the Zone of the Niesen as far as Ormont-Sepey in the Grande Eau valley.

In the Prealps of the Chablais, on the northern side of the Dents du Midi, the Zone of the Niesen does not exist. There, the Median Prealps rest directly on the " *Molasse rouge* " of the Val d'Illiez. The Median Prealps soon disappear to the west, having been laminated by the Brèche Nappe.

At the Col de Golèze, we find slices of the Median Prealps, made up of Lias or Upper Cretaceous, intercalated between the Trias of the Brèche Nappe and the Flysch of the High Calcareous Alps (Fig. 70).

Tectonics. The nappe of the Median Prealps contains several anticlines and synclines often of the type of those of the Jura mountains. Anticlines, however, do not always form the mountain ranges, for these may be carved out of synclines by erosion, as is the case for the Rochers de Naye, above Montreux. The following folds are met with in the Romande Prealps, viz., between the Lake of Geneva and the Lake of Thun :

1. *The external anticline zone,* often called the Lias zone of the Prealps owing to the great rôle played by this formation. Isoclinal and " *schuppen* " structures are the rule, and no arch-bends are visible. This zone begins at Montreux-Chillon, on the Lake of Geneva. Several summits belong to it, as Cubly, Dent de Jaman, Verreaux, Cape au Moine, Dent de Lys, Ganterist. The summit of the Moléson has been eroded out of a syncline dividing two anticlines.

(a) *The great Cretaceous syncline zone of the Gruyère,* often divided into secondary synclines. The beginning of this zone is well marked in the Mont Sonchaux-Rochers de Naye ridge, above Montreux.

2. *The anticline Vanil Noir* (2,386 m.)-*Stockhorn* (2,192 m.). This fold is marked on the Lake of Geneva, above Villeneuve, by the anticlinal valley of the Tinière where Triassic rocks are exposed. The Upper Jurassic of the south-east limb of this fold forms the ridge of Mont d'Arvel, near Villeneuve. The quarries seen at the foot of the latter are situated in Lias. In this anticline the Zoophycos Dogger plays an important rôle, so that the name of " Dogger zone " has been given to it.

(b) *The Flysch syncline of Chateau d'Oex.* This syncline is divided into two parts, near the Rhone valley, by :

3. *The anticline of the Tours d'Aï* (2,334 m.), which soon dies away in the Flysch, to the north-east. The south-east part of the Chateau d'Oex syncline forms :

(c) *The Leysin syncline,* which is marked in the scenery by the great ledge on which the village of Leysin is built.

4. *The Gastlosen range* represents an anticlinal zone thrust over the Leysin syncline and the Chateau d'Oex syncline. It runs from the Rhone valley to the northern side of the Simme valley, through the Laitmaire (see Fig. 71), Dent de Ruth (2,244 m.) and Gastlosen.

(d) *The Flysch syncline of Rougemont-Boltigen-Lower Sim-mental.* This tectonic element terminates the nappe of the Median Prealps to the south-east, where it overlaps the Niesen Flysch. The syncline of Rougemont, like the

Flysch syncline of Chateau d'Oex, is divided into two parts by :

5. *The anticline of the Rubli and Gummfluh.*

In the Flysch syncline of Chateau d'Oex occur remnants of the highest nappe of the Prealps, namely of the Simme Nappe. They float on the Flysch, as will be seen later on. The Rougemont-Lower Simmental syncline also contains remnants of both the Brèche Nappe and the Simme Nappe.

THE BRÈCHE NAPPE

From a tectonic point of view, the nappe of the Median Prealps is overridden by a higher nappe, known as the " Brèche Nappe," owing to the great development of breccias in it.

Stratigraphy. Lugeon has drawn up the following classification of the rocks forming the Brèche Nappe, in descending order :

10. Shales with Fucoids (Flysch) .	Tertiary
9. Marly limestones with Foraminifera (Simme valley) . .	Upper Cretaceous
8. Upper Breccia . . .	Upper Jurassic
7. Slates	Middle Jurassic
6. Lower Breccia, passing horizontally to shales . . .	Middle-Upper Lias
5. Echinodermic limestones (Col de Coux)	Hettangian
4. Lumachelle	Rhaetic
3. ⎧Dolomitic limestones and breccias with gypsum . . ⎫ ⎨Black Clays . . . . ⎬ ⎩Quartzites . . . . ⎭	Trias
2. Conglomerates very like the Verrucano	Permian
1. Micaceous sandstones and slates with anthracite . . .	Middle-Upper Carboniferous

Let us now examine the Jurassic, which is so characteristic, owing to the great rôle played by the breccias.

The shales (Schistes [1] inférieurs of Lugeon) contain carbonate

---

[1] The word " schists " is used by Swiss and French geologists for shales, while British geologists only use it for crystalline formations.

of lime and intercalations of limestones and breccias. When the Rhaetic is exposed, and it is seldom met with, the slates rest on it as is the case at the outcrop of Petit Jutteninge, near Verchaix in the Giffre valley (Haute Savoie). The thickness of the shales may vary from 500 to 1,300 metres in the mountain of the Hautforts (Chablais Prealps).

The facies of the shales occurs in the southern and south-eastern part of the Brèche Nappe. To the north, the shales pass gradually and horizontally to the :

Lower Breccia. The pebbles of the breccia belong mostly to the dolomitic limestones of the Trias. In several cases intercalated beds of echinodermic limestones are met with.

The Lower Breccia rests unconformably on the Trias at the foot of the Pointe de Marcelly, in the Giffre valley (Chablais Prealps), as recorded by Lugeon. The thickness of the Lower Breccia may reach 1,300 metres at the Pointe de Marcelly. Fossils have been found by Lugeon, but none of them permits us to determine exactly the age of the formation.

The slates generally contain red and green clays. The thickness of the whole formation varies from 100 to 300 metres.

The Upper Breccia. It is difficult to distinguish the Upper Breccia from the Lower Breccia. But we may say that the Upper Breccia is characterized by intercalations of compact limestones which are quite different from the echinodermic limestones interstratified in the Lower Breccia. The upper beds of the Upper Breccia contain cherts.

To sum up, *the Upper Breccia contains more limestones than the Lower Breccia.*

The orographic rôle of the formations of the Brèche Nappe. Owing to its great thickness (1,500–2,000 m.) and the hardness of both Lower and Upper Breccias, the Brèche Nappe plays a rôle of paramount importance in the Prealps of the Chablais, which is not the case in the Romande Prealps.

The shales form pastures. The Lower Breccia, on the other hand, is marked in the scenery by steep rocky slopes, or even by great cliffs. The Pointe de Marcelly, in the Giffre valley, is the best example of a peak entirely made up of this formation. The Hautforts, the highest summit of the Brèche Nappe in the Chablais, also consist of Lower Breccia.

Between the summits made up of Upper Breccia and summits

carved out of Lower Breccia, occurs a zone of passes due to the presence of soft slates.

Tectonics. The Brèche Nappe in the Chablais, between the Lake of Geneva and the Arve river, occurs as a very broad syncline (Fig. 70). The frontal fold of the nappe appears only at the Pointe de Grange (2,438 m.), between the Drance of the Biot valley and the Drance of the Abondance valley.

Between the Lake of Geneva and the Lake of Thun, the Brèche Nappe is known as the " Hornfluh Brèche " from the name of a mountain situated between Gstaad and Zweisimmen. The tectonics of the nappe are more complicated in this part of the Prealps than in the Chablais. Indeed, as shown in Fig. 71, two digitations are met with in its frontal part. They occur on the southern slope of the Flysch syncline of Rougemont belonging to the Median Prealps.

THE SIMME NAPPE

On the Flysch of the Median Prealps, between the Lake of Geneva and the Lake of Thun, occur slices of the highest nappe of the Prealps : the Simme Nappe (Fig. 69).

Some of these slices have been studied by Jeannet and Rabowsky on the Flysch of the Chateau d'Oex syncline, and by Rabowsky in the Flysch of the Rougemont-Lower Simmental syncline.

Rocks. To understand the relations of the Simme Nappe with the Eastern Alps (Austrides), let us study the stratigraphical sequence of this nappe, according to Rabowsky, in descending order :

Upper Cretaceous.	Marly limestones and shales with *Globigerina*. At the base occur intercalations of echinodermic limestones with *Orbitolina* (Cenomanian).
Lower Cretaceous. Upper Jurassic.	White limestones with *Aptychus, Pygope janitor, Simoceras volanenze, S. biruncinatum.*
Middle Jurassic.	Radiolarites.
Upper Lias.	Limestones with brown siliceous shales, containing Ammonites of the Aalenian.

Such a succession of rocks is unknown in the High Calcareous Alps, as well as in the Pennine Nappes.

Beside these sedimentary rocks, patches of eruptive rocks are met with in the Flysch. They are limited to the southern syncline, while in the northern syncline of the Median Prealps there occur only slices of the sedimentary rocks, from the Upper Lias to the Upper Cretaceous. In the last case, there is no doubt that the Simme slices overlap the Flysch. In the southern syncline, on the other hand, Rabowsky has shown that the relations between the slices of the Simme Nappe and the Flysch of the Median Prealps are complicated. In fact, the Simme slices are stretched and often form lenticles penetrating into the Flysch, or are intensely refolded.

The eruptive rocks, which accompany the sedimentary rocks, consist of gabbros, ophites, porphyrites. Similar formations occur in the Flysch of the broad syncline of the Brèche Nappe at Les Gets, above Taninge, in the Chablais Prealps. As shown by Michel Lévy, the following rocks are met with : granite, serpentine, diabases and gabbros, porphyrites, variolite, kersantite. A patch of this last rock, as shown by Lugeon, occurs at Les Farquets, near Somman, *on the thrust plane of the Brèche Nappe.*

Tectonics. When comparing the very simple tectonics of the Brèche Nappe with the complicated structure of the Simme Nappe, Rabowsky expressed the opinion that the Simme Nappe was not the highest element of the Prealps, this rôle being played by the Brèche Nappe.

Owing to the travelling of the very thick Brèche Nappe, the Simme Nappe was stretched, dragged and accumulated on the frontal part of the Brèche Nappe. The patch of kersantite of Les Farquets, situated on the thrust plane of the Brèche Nappe, would confirm this hypothesis.

It is easy to reply to Rabowsky, as several authors did, that if patches of the Simme slices, in rare cases, occur on the thrust plane of the Brèche Nappe, near its frontal part, we have to deal with an involution of the Simme Nappe between the Brèche Nappe and the Median Prealps. This phenomenon of involution may also explain the complicated structure of the Simme Nappe, in front of the Brèche Nappe.

To sum up, *the Simme Nappe belongs to the Austrides and must be considered as the highest element of the Prealps.*

The Klippes (Outliers)

The Klippes (outliers) of the Prealps permit us to determine the extent of the nappes of the Prealps. They occur in synclines of the High Calcareous Alps, where they have been spared by erosion.

Their substratum may consist of the Flysch of the Wildhorn Nappe, or of the Wildflysch of the External and Internal Prealps (upper nappes of the High Calcareous Alps).

The Klippes extend from near the Lake of Annecy, on the south-west, as far as the Rhine valley, on the north-east.

The following are the Klippes, from south-west to north-east :

1. *Sulens*, situated in the south-western part of the Serraval-le-Reposoir syncline, separating the Autochthon from the Morcles Nappe.

2. *The Annes*, in the north-eastern part of the syncline referred to above. As shown by Moret, both the Klippes of Sulens and of the Annes show remnants of nappes belonging to the Median Prealps, the Niesen Flysch and the External Prealps.

3. *The Klippes of the Chatillon Plateau*, between the Giffre valley and the Arve valley. Moret has shown that several small Klippes are met with in this region. The *Saint Sigismond* Klippe, 2 kilometres in length, consists of formations belonging to the Brèche Nappe. The *Chatillon* Klippe represents the only known remnant of the Niesen Flysch in this region. At Samoens, on the left bank of the Giffre, Collet and Beck discovered a Klippe belonging to the Wildflysch of the Internal Prealps.

4. *The Giswilerstöcke*, between the Lake of Brienz and the Lake of Lucerne, represent a complicated Klippe made up of three slices belonging to the Median Prealps.

5. *The Stanserhorn*, on the south-western side of the Lake of Lucerne, accompanied by the *Arvigrat, Buochserhorn, Klevenalp, Musenalp*. All these Klippes consist of remnants of the Median Prealps.

6. *The Mythen*, above Brunnen and Schwyz (see Fig. 72), are remnants of the Median Prealps, resting on the Flysch of the Säntis Nappe.

7. *The Iberg* Klippes, situated between the Mythen and the Sihl valley, consist partly of formations belonging to the Simme Nappe and partly to the Median Prealps. They do not play any topographical rôle, like the other Klippes referred to above.

8. *The Grabs* Klippe, in the Rhine valley, is made up of formations which show more likeness to the rocks of the

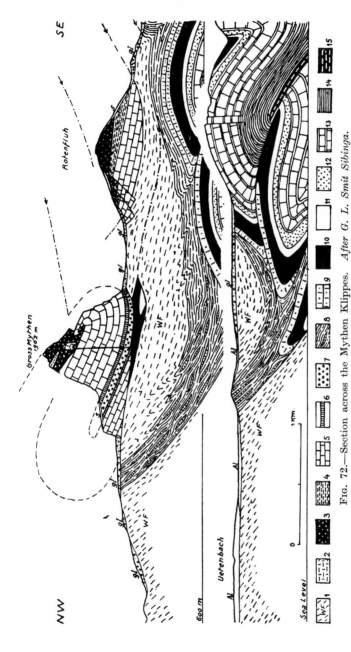

FIG. 72.—Section across the Mythen Klippes. *After G. L. Smit Sibinga.*

Klippes.—1, Wildflysch. 2, Upper Cretaceous (grey limestones). 3, Upper Cretaceous (red and green shales). 4, Lower Cretaceous. 5, Upper Jurassic. 6, Dogger. 7, Trias.

Drusberg Nappe (High Calcareous Alps).—8, Flysch. 9, Nummulitic limestones. 10, Upper Cretaceous ("Seewen marls"). 11, Upper Cretaceous ("Seewen limestones"). 12, Gault. 13, Barremian. 14, Hauterivian. 15, Valanginian. *Al,* Alluvial matter. *S,* Screes. *gl,* Moraines.

Falknis, in the Eastern Alps, than to the other Klippes of the Prealps. We will come back to this problem when dealing with the relations between the Prealps and the Austrides.

THE GIFFRE VALLEY IN THE CHABLAIS PREALPS

A good natural section of the Chablais Prealps is exposed in the Giffre valley, which may be easily reached from Geneva by motor-car. Moreover, this excursion may be completed in the same day, in crossing the High Calcareous Alps, along the Arve valley, as shown on page 79.

From Geneva to Annemasse we see the Salève anticline disappearing under the soft Aquitanian sandstones (Molasse) owing to its pitching towards the north-east.

From Annemasse we have in front of us the Voirons belonging to the *External Prealps* (Jurassic, Cretaceous and Tertiary). At Bonne we realize that the valley which we follow is an old channel of an important river that has been captured by an affluent of the River Arve. We shall see, later on, the place where the capture has taken place. On our right we have a low ridge on the front of the Môle, the Collines de Faucigny, that forms the prolongation of the *External Prealps*. At Viuz we enter the *Median Prealps* (Trias to Tertiary), the Trias of which is resting on the Tertiary of the External Prealps. The Môle and the Brasses belong to the Median Prealps.

At Pont du Risse we notice a small stream flowing towards the Alps, i.e., towards the capture referred to above. Within a few minutes we see the River Giffre changing its direction through a right angle to flow into the River Arve; it is the place where the capture begins.

From Mieussy, the Pointe de Marcelly appears, forming an important cliff above the valley. It represents a new tectonic element—a higher nappe, the *Brèche Nappe*, resting on the Median Prealps. The stratigraphical sequence of the Brèche Nappe begins with Carboniferous sandstones and ends with Tertiary. Thick beds of breccias occur in the upper part of the cliff of the Pointe de Marcelly (Fig. 70).

From Taninges we have to cross the Col de Chatillon to arrive in the Arve valley. To the north-west of the Col appears the Mont Orchex (Median Prealps) resting on the *Niesen Flysch* (Nummulites) in which the Col has been eroded. Looking towards the north-east and east we get a good view of

PLATE X.   The Mythen, a Klippe (Outlier) of the Prealps, resting on the Tertiary of the High Calcareous Alps.

Photo by Wehrli-Verlag.

the relation between the High Calcareous Alps and the Prealps, as shown in Fig. 70.

During the descent to the Arve valley we enjoy the scenery of the High Calcareous Alps, at the foot of which the town of Cluses is built.

CHAPTER II

CORRELATIONS BETWEEN THE PREALPS AND OTHER ALPINE NAPPES

R. Staub, followed by the Swiss geologists who mapped in the Grisons (north-eastern part of the Swiss Alps), correlates the three upper nappes of the Prealps to the Austrides.

Lately two French geologists, Gignoux and Moret, on the basis of Haug's working hypothesis, and researches in the field, put forward the view that the three upper nappes of the Prealps root in the Pennine Nappes. Let us summarize these interpretations as follows :

Succession of Nappes.	Staub's Correlation.	Gignoux and Moret's Correlation.
Simme Nappe . . .	Campo or Silvretta Nappe	Dent Blanche Nappe, VI
Brèche Nappe . . .	Bernina Nappe	Great St. Bernard Nappe, IV
Median Prealps Nappe	Err Nappe	
Niesen Flysch Nappe.	Dent Blanche Nappe, VI	Simplon Nappes, I–III

External and Internal Prealps .   . High Calcareous Alps

In the *Guide géologique de la Suisse* (1934) Gagnebin discusses the correlations established between the Prealps and the other Alpine Nappes. He points out that, at present, we may consider as demonstrated that : 1. The Niesen Flysch belongs to the Middle digitations of the Great St. Bernard Nappe. 2. The Simme Nappe must be connected with the lower Austrides (Err Nappe or Bernina Nappe). Thus we arrive at the following results :

Simme Nappe	Lower Austrides (Err Nappe or Bernina Nappe).
Brèche Nappe	Not represented in either Pennine or Austride Nappes.
Median Prealps Nappe . .	An Upper Pennine Nappe.
Niesen Flysch Nappe . .	Great St. Bernard Nappe.
External and Internal Prealps .	Upper Nappes of the High Calcareous Alps.

CHAPTER III

THE AGE OF THE TRAVELLING OF THE PREALPS

When dealing with the Swiss Plateau (p. 118), we have seen this Kinetic picture : *The Alps travelling forward, and the pebbles going forward from the chain as it grows and the chain riding over its own debris.* Pebbles coming from the disintegration of the Prealps are to be found in the conglomerates of the Mont Pélerin, above Vevey, on the northern side of the Lake of Geneva. This formation of Chattian age (Middle Oligocene) demonstrates that the Prealps had already travelled to the north at this time. But their advance over their own debris is of *Upper Miocene age*, as demonstrated by the fact that the Marine Molasse (Middle Miocene) of the Swiss Plateau is overridden by Middle Oligocene conglomerates over which the Prealps have been thrust.

BIBLIOGRAPHY

Only fundamental references are given here. Earlier literature will be found mentioned in them.

1. SCHARDT, H.—Sur l'origine des Préalpes romandes (zone du Chablais et du Stockhorn). Arch. Sc. phys. et nat. Genève, 3 période, T. xxx, pp. 1–14, 1893.
2. SCHARDT, H.—Les régions éxotiques du versant nord des Alpes Suisses. Bull. Soc. vaud. des Sc. nat., T. xxxiv, p. 113.
3. LUGEON, M.—Les grandes nappes de recouvrement des Alpes du Chablais et de la Suisse. Bull. Soc. géol. de France, iv sér., vol. i, pp. 723–825, 1901.
4. JEANNET, A.—Das romanische Deckengebirge, Préalpes und Klippen in Geologie der Schweiz, by Albert Heim, pp. 589–676, Bd. ii, 1922.
5. GAGNEBIN, E.—Les Préalpes et les " Klippes." Guide Géologique de la Suisse, pp. 79–95. Wepf et Cie, Bâle, 1934.

PART VI

THE ALPINE RANGE IN THE WESTERN MEDITERRANEAN

CHAPTER I

INTRODUCTION

Whilst the Alps *sensu stricto* extend from Genoa to Vienna, the Alpine Chain, as shown by Eduard Suess, can be followed geologically into the Western as well as the Eastern Mediterranean Region. Indeed, the geology of the Alpine Chain cannot be separated from the geology of the Mediterranean Region.

Argand, R. Staub and Kober have dealt with this important question in recent publications. W. von Seidlitz in his large book, published in 1931, *Diskordanz und Orogenese der Gebirge am Mittelmeer*, attempted a synthesis on this fascinating subject.

Since the exploration and mapping in different parts of the Mediterranean Region vary considerably in standard, all these authors have been led into some degree of speculation. Much more is known of the western part of the Mediterranean Region than of the Balkans.

In this new part we shall not deal with the geological history of the Mediterranean Region but simply try to follow the Alpine structures in the Western Mediterranean. I have studied in this region many of the facts which will allow me to give my own opinion when there are conflicting theories to be considered.

CHAPTER II

THE ALPS AND APENNINES

1. WHERE DO THE ALPS EXTEND BEYOND GENOA ?

As demonstrated by Termier, along the Mediterranean shore from Nice to Albenga one crosses the Maritime Alps which are the sedimentary cover of the Hercynian Massif of Mercantour. One is therefore on the Foreland.

At Albenga the formations of the Alpine geosyncline appear represented by the Great St. Bernard Nappe, and farther east, at Savona, by the Dent Blanche Nappe. Eastwards a zone of *Schistes lustrés* with ophiolites follows and is developed as far as Genoa.

We shall name these *Schistes lustrés*, for the purpose of later discussion, the *Voltri Schistes lustrés*, after a village in the vicinity.

The arc of the Western Alps terminates at the Mediterranean Sea, between Nice and Genoa. E. Suess, in 1885, put forward the view that the Apennines and Atlas were a continuation of the Alps towards the south and the west. Then by means of an arc Suess united the Betic Cordillera, of Southern Spain, to the Atlas. This picture was modified by Termier, in 1911. This author regarded the Alps as extending from Genoa to the Rif, Gibraltar and the Betic Cordillera across the Balearic Islands, and the Apennines and Atlas as a separate chain branching off the Alps at Genoa. Between the Alpine Chain and the Apennines-Atlas Chain he supposed the existence of an Hercynian massif of which the western part of Corsica and the whole of Sardinia were the remnants. Fig. 73 shows the modifications that have been introduced in Suess' and Termier's ideas by Kober in 1912, Staub in 1924, Stille in 1927.

At present, we have more definite facts at our disposal to establish the continuation of the Alps in the Western Mediterranean. If we want to follow the Alps in this direction, we must, first of all, make sure that we are in the Alps. How shall we proceed ?

We have to look for the most characteristic rocks of the Alpine geosyncline, i.e., the *Schistes lustrés* and ophiolites (greenstones). Then, as we know that the nappes of the Alpine geosyncline have been thrust over the Foreland we must ascertain that this foreland exists. *Corsica*, as shown by Termier, exemplifies these principles. The western part of this island is made up of a granitic massif of Hercynian age, which acted as Foreland, and very characteristic *Schistes lustrés* occur in several nappes in the eastern part of Corsica. Whilst leaving till later pages the more detailed account of the Alpine structures of Corsica, one may remark here the unanimity of authors who have worked in this island. Argand, Staub, Kober, Seidlitz, and Paréjas, after researches in the field, have all accepted Termier's demonstration based on Maury's mapping. Although all are in agreement that the movement is directed from east to west, Kober and Seidlitz, however

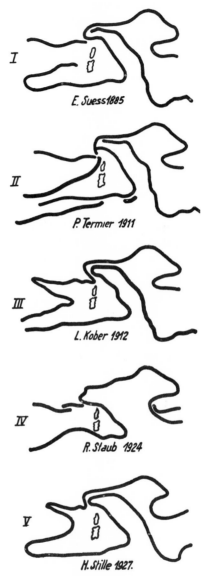

FIG. 73.—The Alpine Chain in the Western Mediterranean.

explain this by the hypothesis that in Corsica one is only concerned with the west wing of their double Alpine thrusting (see page 23).

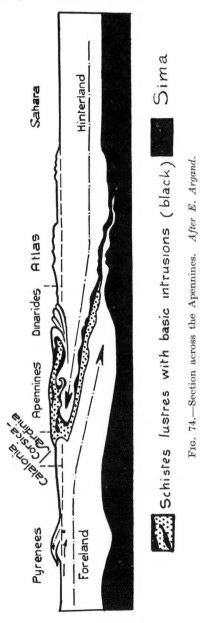

Fig. 74.—Section across the Apennines. *After E. Argand.*

Argand in his *Tectonique de l'Asie* sees in the Northern Apennines : the D i n a r i d e s. But the K l i p p e s (outliers) of ophiolites and radiolarites, discovered by Steinmann and Tillmann in the provinces of Liguria and Tuscany, represent for Argand a nappe formed in the Alpine geosyncline which has been thrust over the Dinarides, by means of a gigantic backward folding (Fig. 74). According to Argand this nappe belongs only geographically to the Apennines. We shall see in the next chapter that R. Staub does not accept this view. For him the Klippes of Liguria and Tuscany belong to a more internal tectonic element ; they have nothing to do with the Pennine Nappes of the Alps, and the backward folding advocated by Argand does not exist (Fig. 75).

Moreover, we shall see in the next chapter that, owing to the existence of the window of the Alpi Apuane in Tuscany, the Northern Apennines belong also to the Alps.

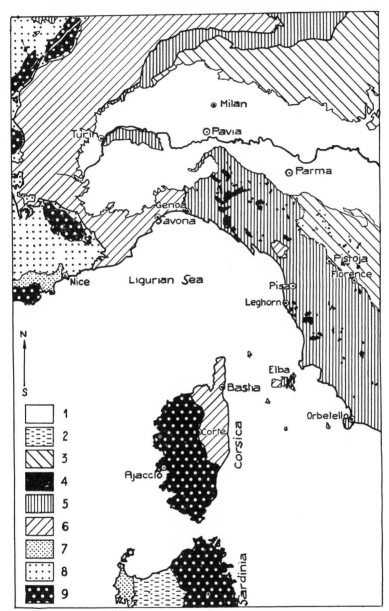

FIG. 75.—Tectonic Map of the Northern Apennines and Corsica.
After R. Staub.

1, The plain of River Po. 2, The volcanic region of Sardinia. 3, Dinarides. 4, Klippes (outliers) of the Liguride Nappe. 5, Austrides (Spezzia Nappe). 6, The Pennine Nappes. 7, Sedimentary of Provence and Sardinia. 8, Sedimentary of the High Calcareous Alps. 9, Hercynian Massifs.

CONCLUSIONS. The Western Alps geographically terminate between Nice and Genoa, on the Mediterranean Sea. Geologically they reappear in Corsica, where we find a part of their Foreland and nappes formed in the geosyncline, and in the Northern Apennines.

2. THE WINDOW OF THE ALPI APUANE

Introduction. On following the railway line along the western coast of Italy, between Genoa and Leghorn (Livorno), the soft scenery of the hills of the Ligurian Apennines is seen to change at Sarzana and between this locality and Pisa appear on the left the *Alpi Apuane*, better known to Englishmen as the *Carrara mountains*. Douglas Freshfield gives a vivid description of this region : " Above the waters of the soft blue Spezzian bay they thrust themselves out of an unchanging belt of grey olive woods, a cluster of keen swordblades, in winter white with alpine snows, in summer golden in the sunshine with the warm hues of weathered marble, purple in the shade with the rich bloom of a southern atmosphere."

The highest crags rise nearly 2,000 metres directly above the sea. A solitary tower stands at either end. The northern is the Monte Sagro, the southern the Pania della Croce. For the Alpine geologist who has been wandering along the dull hills of the Apennines, this cluster of peaks has a great fascination.

In autumn when the geologist is driven out of the Alps by fresh snow the field work of the year dies away. In the Alpi Apuane and in the Apennines he may spin it out, for there autumn is a riper summer.

If Ruskin made the Alpi Apuane the subject of one of his eloquent panegyrics, Zaccagna, an Italian geologist, made them his life work. On the base of Zaccagna's detailed mapping Lencewicz discovered two different nappes. In 1926 Tilmann, a German geologist, revised and enlarged the work done by Lencewicz and pointed out that the Alpi Apuane are made up of three nappes of the Alps, in his opinion one belonging to the Pennine Nappes, the second to the Austrides and the third to the Dinarides. But the results arrived at by Tilmann were puzzling, for his sequence of nappes was the reverse of what happens in the Alps, i.e., his Pennine Nappe was on the top and the Dinarides on the bottom. To get

out of this difficulty the great backward folding was con-
structed by Argand, as already shown (page 258). Lately
R. Staub went to the field to check Tilmann's views and now
proposes a new interpretation of the tectonics of the Alpi
Apuane, which, from what I have seen in the field, corresponds
much better with the facts.

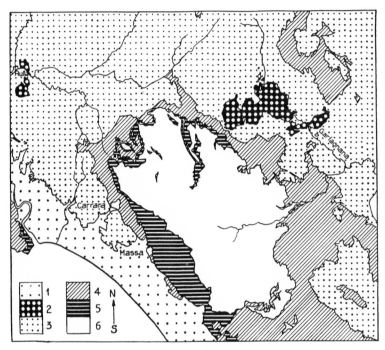

FIG. 76.—The Window of the Alpi Apuane.   *After R. Staub* (simplified).

1, Coastal plain.   2, Klippes (outliers) of the Liguride Nappe.   3, Tertiary of the Spezzia
Nappe.   4, Mesozoic of the Spezzia Nappe.   5, Massa belt.   6, The Carrara Nappe.

The Nappes. R. Staub sees in the Alpi Apuane the
following three nappes in descending order :

3. The Liguride Nappe.
2. The Spezzia Nappe.
1. The Carrara Nappe.

For Staub the Alpi Apuane are a *window*, hence their importance for the study of the relations between the Alps and
Apennines. This window is due to erosion on an axis elevation.

The Carrara Nappe, the lowest tectonic element, is located

in the window, the frame of which is represented by the Spezzia Nappe. The Liguride Nappe consists only of Klippes (outliers) resting on the Tertiary of the Spezzia Nappe (Fig. 77).

THE CARRARA NAPPE

According to R. Staub, the stratigraphical sequence is the following, from top to bottom :

10. Tertiary.—Flysch, sandstones called Pseudomacigno.
9. Upper Jurassic.—Radiolarian clays of a green, red or violet colour.
8. Upper Jurassic.—Radiolarites, red or green.
7. Middle Jurassic.—*Schistes lustrés*, with breccias of the Dolin (page 180) type. Fuchsite occurs in them and at the top these *Schistes lustrés* pass into the radiolarites.
6. Lias.—Carrara marbles, with beds of breccias.
5. Rhetian.—Banded limestones separating the Triassic formations from the Carrara marbles.
4. Trias, of the type of the lower Austrides, called Grezzoni.
3. Permian.—Verrucano made up of breccias, arkoses and conglomerates.
2. Carboniferous.—Black and green phyllites, brown sandstones.
1. Crystallines.—Gneiss, micaschists, phyllites, green schists and amphibolites.

The upper part of this series (Nos. 6–10) represents the Schams slices and the Platta zone of the upper part of the Dent Blanche Nappe (see page 183). The Carrara Nappe thus belongs to the upper part of the Pennine Nappes of the Alps.

The Triassic rocks and the Carrara marbles play a prominent rôle in the scenery of the Alpi Apuane, for several peaks have been carved out of them.

The Carrara Nappe, as shown in the tectonic map (Fig. 76), occupies the central part of the window of the Alpi Apuane. In the region of Massa the Carrara Nappe is separated from the Spezzia Nappe by a series of slices named the *Massa zone* which extends along the north-west margin of the Carrara Nappe.

FIG. 77.—Section across the Window of the Alpi Apuane. *After R. Staub.*

1, Klippes (outliers) of the Liguride Nappe. 2, Tertiary of the Spezzia Nappe. 3, Mesozoic of the Spezzia Nappe. 4, Crystallines of the Spezzia Nappe. 5, Massa Belt. 6, Mesozoic of the Carrara Nappe. 7, Trias and Marbles of the Carrara Nappe. 8, Crystallines of the Carrara Nappe.

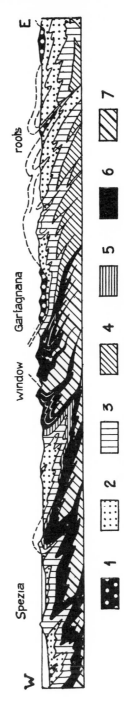

FIG. 78.—Section across the Northern Apennines. *After R. Staub.*

1, Klippes (outliers) of the Liguride Nappe. 2, Tertiary of the Spezzia Nappe. 3, Mesozoic of the Spezzia Nappe. 4, Crystallines of the Spezzia Nappe. 5, Massa Belt. 6, Mesozoic of the Carrara Nappe. 7, Crystallines of the Carrara Nappe.

263

The Spezzia Nappe

The Spezzia Nappe has been thrust over the Carrara Nappe, but at present forms the frame of the Alpi Apuane window, having been eroded out on the axis culmination.

The normal sequence of the Spezzia Nappe is *the normal sequence of the Northern Apennines* in Liguria and Tuscany. This is of paramount importance for the study of the relations between the Alps and Apennines.

9. Tertiary.—Nummulitic limestones and Macigno.
8. Cretaceous.—Marly limestones at the base and red clays of deep-sea facies in the upper part, named Scaglia.
7. Upper Jurassic.—Radiolarites.
6. Middle Jurassic and Lias.—Very fine limestones of deep sea facies.
5. Rhetian.—Fossiliferous beds.
4. Trias.—Quartzites and dolomitic limestones.
3. Permian.—Verrucano in Monte Pisano.
2. Carboniferous, at Jano.
1. Crystallines.

In opposition to the Carrara Nappe, the rocks of the Spezzia Nappe show no metamorphism at all. Moreover, they do not contain any ophiolites. Though R. Staub points out that we have to deal, in the Spezzia Nappe, with a stratigraphical series which reminds us of the lowermost part of the Grisonides (lower Austrides), he thinks that it might also represent the topmost part of the Dent Blanche Nappe, i.e., the uppermost part of the Pennine Nappes of the Alps.

According to Staub the Northern Apennines, to the southwest of a line indicated by the following localities : Orvieto, Arezzo, Florence, Pistoja, Parma, are made up of the Spezzia Nappe. It is generally the Tertiary of this latter nappe that crops out at the surface of the ground, the Cretaceous and Jurassic being only seen in natural cuttings due to erosion. The **bottom** part of the nappe occurs only on the border of the window of the Alpi Apuane and near Genoa.

The outliers of ophiolites and radiolarites which are scattered all over the Tertiary of the Spezzia Nappe represent the remnants of a higher nappe : *the Liguride Nappe.*

The Spezzia Nappe being an element of the Austrides covering a Pennine nappe, as shown in the window of the

Alpi Apuane, *the Northern Apennines from Genoa to Orbetello belong, geologically speaking, to the Alps* (Fig. 75).

THE LIGURIDE NAPPE

We owe our knowledge of the Liguride Nappe to Steinmann. During many years this author studied the interesting rocks which are found as isolated caps on the Tertiary of the Ligurian and Tuscan Apennines (Spezia Nappe). These rocks had been considered by the Italian geologists to represent the upper part of the Tertiary, and ophiolites (greenstones) being the most characteristic of these rock types, they were termed " Ophiolitic Tertiary " (*Terciario ofiolitico*).

The ophiolites are always accompanied by red radiolarites and clays, and also by white or grey sublithographic limestones (*Alberese* of the Italian geologists).

On the evidence of microscopical studies of the red clays and radiolarites Steinmann pointed out that these rocks are of an abyssal facies, i.e., are equivalent to the abyssal red clays of the present Ocean. In a normal series the radiolarites and red clays are overlain by the sublithographic limestones.

The sublithographic limestones contain foraminifera, of which *Calpionella alpina* Lor. is characteristic of the Upper Jurassic and Lower Cretaceous. Hence the radiolarites are older than the Upper Jurassic and Steinmann pointed out that these formations must belong to a nappe thrust over the Tertiary of the Apennines.

The ophiolites of the Ligurian and Tuscan Apennines consist of serpentines, gabbros, diabases and spillites. Amongst these, serpentines are predominant. Granite sometimes occurs with the ophiolites. Some authors, not accustomed to Alpine geology thought that this granite had intruded the ophiolites, but Steinmann expressed the view that one had to deal merely with *slices* peeled off the substratum by the overriding nappe. Lately the boring of a tunnel in the Garfagnana, a valley to the north of the Alpi Apuane, has shown that Steinmann's views were right. The reader who would like to get more particulars on this question should consult the paper by G. Merla, a young Italian geologist, listed in the Bibliography.

What are the relations between the above-mentioned sedimentaries and the ophiolites ? The ophiolites cap the sedimentaries or occur in great lenticles in them. Furthermore, pillow lavas being frequent in the ophiolites, we may conclude that the basic magma has been forced for great distances between the sedimentaries, forming sheets, or has escaped at certain places into the bottom of the sea.

The age of intrusion is difficult if not impossible to determine. Indeed, the stratigraphical relations in the Liguride Nappe have been altered by movements within the nappe itself. The ophiolites being more massive than the sedimentaries have slid sometimes on, sometimes in the latter. As a result crush breccias and subsidiary folding occur.

The ophiolites with their sedimentaries do not form a complete nappe, they are only klippes (outliers) which often break the monotony of the scenery of the Ligurian and Tuscan Apennines.

3. THE DIRECTION OF MOVEMENT IN THE NORTHERN APENNINES

To accept Staub's views of a travelling westwards of the nappes of the Northern Apennines, we must know if there is any evidence of a belt of roots. According to R. Staub's map (Fig. 75), the boundary between the Alps and Dinarides occurs along a line marked by the following localities : Orvieto,

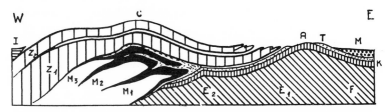

FIG. 79.—Section across the Northern Apennines. *After Kober.*

1, "Betwixt mountain" near Elba. *C*, Culmination of Carrara. *Z2*, Liguride Nappe. *Z1*, Toscanides (Spezzia Nappe). *M1, M2, M3*, Metamorphides (Pennine Nappes). *A*, Autochthon of the Abruzzi. *E1, E2*, Externides. *F*, Foreland. *K*, Mesozoic. *T*, Flysch. *M*, Molasse.

Arezzo, Florence, Pistoja, and Parma.   In Fig. 78 we see the roots of the Ligurides Nappe intercalated between the Alps and Dinarides.   But outliers of the Liguride Nappe occur not only to the west of the belt of roots, but also to the east of it.   To the west the outliers, made up of ophiolites and radiolarites, rest on the Spezzia Nappe, to the east outliers of the same composition are to be found *on the Dinarides.*   The

Liguride Nappe thus comes out of a deep-sea basin intercalated between the Austride region and the Dinaric region.

In the making of the Alps, the underthrusting due to the formation of the Dinarides (Plate I, Fig. 13) induced a fan arrangement of the Liguride Nappe, i.e., an outward over-thrusting towards the Alps and towards the Dinarides.

These are the views exposed by R. Staub. Let us now examine the ideas developed by Kober on the same subject.

According to Kober, the Northern Apennines belong to the Dinarides and not to the Alps. The nappes have travelled eastwards and not westwards, as proposed by Staub. To understand Kober's statement we must anticipate what we shall see in Corsica. All the authors, including Kober, who have been dealing with this island admit that the Alpine Nappes there have travelled westwards. Thus Kober admits westward-moving nappes in Corsica and eastward-moving nappes in the Northern Apennines. This is quite understand-able if we accept, with Kober, the existence of " betwixt-mountains " located near or in the island of Elba.

Evidence to accord with this hypothesis is, however, extremely scanty. Indeed, if we compare Kober's and Staub's sections (Fig. 79 and 78) with the evidence on the field, we arrive at the result that we much prefer a westward movement of the nappes in the Northern Apennines, as advocated by Staub.

4. CONCLUSIONS

To understand the relations between the Apennines and the Alps we must not forget that the Tertiary of the Northern Apennines is the Tertiary of the Spezzia Nappe, as demon-strated by the window of the Alpi Apuane (Fig. 77).

Let us now consider the relations between the Spezzia Nappe and the *Voltri Schistes lustrés*.

Near Genoa, on its western side, between Sestri Ponente and Voltaggio, one sees the *Voltri Schistes lustrés*, very much crushed, overridden by the Spezzia Nappe. Indeed, the dolomitic Trias which represents the base of this latter nappe is thrust over the *Voltri schistes lustrés* as shown in several localities such as : Madonna del Gazzo, Voltaggio and San Pier de Pré.

The *Voltri schistes lustrés* represent the sedimentary cover of the Dent Blanche Nappe, which crops out at Savona.

On the other hand, the Spezzia Nappe, according to Staub, is equivalent to the Austrides. Thus, near Genoa, at the very place where geographically the Northern Apennines unite to the Alps, we see only a boundary between two Alpine nappes : a Pennine (Dent Blanche) and one of the Austrides.

The window of the Alpi Apuane showing in its centre the reappearance of a Pennine nappe (the Carrara Nappe) and being framed by an element of the Austrides (the Spezzia Nappe), we arrive at the conclusion that the *Northern Apennines between Genoa and Orbetello are the continuation of the Alps.* The Liguride Nappe, represented only by outliers of ophiolites and radiolarites, belongs to the Dinarides. As shown in the map (Fig. 75) the boundary between the Alps and Dinarides crosses the Apennines on a line marked by the following localities : Orvieto, Arezzo, Florence, Pistoja and Parma.

CHAPTER III

CORSICA

INTRODUCTION

Corsica is a great key to the study of the Alpine Chain in the western Mediterranean.

From a geological point of view, Corsica can be divided into two regions which present well-marked contrasts in Scenery :

1. A *granitic region*, forming the western and southern part of the island : *Hercynian Corsica*.

2. A *sedimentary region*, represented in the north-eastern part of the island : *Alpine Corsica*.

To Haug we owe the first idea of the rôle that Corsica might play in Alpine geology, for, as far back as 1896, he formulated an hypothesis in which the schists and greenstones of eastern Corsica were regarded as the continuation of the *Schistes lustrés* of the Alps.

In 1905, Termier, after field researches, based on Haug's working hypothesis, *demonstrated* that the schists and greenstones of eastern Corsica are the continuation of the *Voltri Schistes lustrés* (page 256), which have been deposited in the Alpine geosyncline. This discovery is of paramount importance in proving that Corsica is a part of the Alpine Chain.

A few years later Termier published his view that the Alpine Nappes of Corsica, of which he recognized only two at the time, had travelled from east to west and had been thrust over the granitic region. Corsica thus is the *geological continuation of the Western Alps which, in the geographical sense, terminate between Nice and Genoa.*

GRANITIC CORSICA

Granitic Corsica together with Sardinia,[1] represent the remnants of an Hercynian granitic massif, which is the

---

[1] See Fig. 75.

equivalent of the Hercynian massifs of the Foreland of the Alps, like Maures-Esterel, Mercantour, Pelvoux-Belledonne, Mont-Blanc-Aiguilles Rouges, Aar-Gastern. In Corsica, at places where the granite, named protogine by Nentien in 1897, has been overridden and crushed by the Alpine Nappes, mylonites are developed. Deprat, in 1905, drew attention to these modifications of the granite, and Termier pointed out that the granite may be transformed into an *ultramylonite*, i.e., a compact greenish rock, which macroscopically quite unlike a granite, under the microscope is seen to be an extremely mashed granite.

A general section, from west to east, at the contact between Hercynian Corsica and Alpine Corsica shows the following structural elements :

1. The Hercynian granite, sometimes pitching strongly to the east.

2. The normal sedimentary cover of the granitic massif, made up of Mesozoic limestones and Tertiary sandstones and conglomerates.

3. A zone of crystalline slices due to the breaking of the eastern border of the granitic massif.

4. The frontal part of the nappes which were developed in the Alpine geosyncline.

Paréjas, who studied carefully this section in the field, has shown that the rocks of the sedimentary cover of the granitic massif (No. 2) are all dynamometamorphosed and that these rocks being of Mesozoic and Tertiary age, only Alpine folding can account for the phenomenon. To the south of Corte, Paréjas found blue limestones at the base of the sedimentary cover of the granite transformed into marbles, while the same rocks are less dynamometamorphosed to the north of this locality. It is worth noting that these blue, fine-grained, limestones show a great resemblance to the Malm of the High Calcareous Alps (Fig. 16) of Switzerland.

Examining the response of the Mont-Blanc Massif to the Alpine folding (page 44) we have seen that this rigid massif could not fold, but yielded by breaking up into a series of slices, often wedge-shaped. In the same manner the eastern border of the Hercynian Massif of Corsica, which was exposed to the push of the Alpine Nappes, broke into wedges separated from one another by sedimentaries of their former cover.

This structure demonstrates, without any doubt, that the travelling of the Alpine Nappes in Corsica was directed from east to west and that the granitic massif is really equivalent to the Hercynian massifs of the Foreland of the Alps.

At first, Steinmann, Stille and Kober, considering the granitic massif as a " betwixt-mountain " (Zwischengebirge) postulated that the nappes had travelled *from* it towards the east, viz., towards the Apennines. At present, Kober and Seidlitz accept also Termier's views of a movement of the nappes *towards* the granitic massif, but with the restriction that we have only to deal with the westward-moving part of Kober's double thrusting. In such a case, Kober's " *betwixt-mountain* " is not the granitic part of Corsica but must be located somewhere between Corsica and Elba, where it is now at the bottom of the sea.

Argand and Staub have accepted Termier's views that granitic Corsica is an equivalent to the Hercynian massifs of the Foreland of the Alps.

ALPINE CORSICA

In 1928 Termier, after an excursion in Corsica with Steinmann, Kober, Staub, Tilmann and Raguin, considered Alpine Corsica to be made up of the following nappes, in descending order :

3. Outliers of an upper nappe, representing part of the Austrides.
2. Outliers of a nappe of ophiolites and radiolarites, equivalent to the very lowest part of the Austrides.
1. A nappe of *Schistes lustrés*, with ophiolites and radiolarites, corresponding to. the Pennine Nappes of the Alps.

Staub, after the excursion with Termier in Corsica, also published a paper on the nappes of this island. but in much greater detail.

It is impossible, without a detailed geological map, to review this publication here. It is quite sufficient for us to know Staub's results. In a general way Staub's succession of the nappes is identical with that of Termier, but he endeavours to be more precise in correlating with the nappes of the Alps. Nappe 1 is for Staub the equivalent of the Great St. Bernard and Monte Rosa Nappes. In nappe 2, he sees the Dent Blanche, and nappe 3, is equivalent to the Austrides.

It is important to remark that when Staub tries to correlate the nappes of Corsica with those of the Ligurian and Tuscan Apennines, he asks himself if nappe 2 of Corsica, which is characterized by the ophiolites and radiolarites, should not be correlated with the Liguride Nappe. This question seems very important indeed, since a French geologist, Jodot, has demonstrated that some limestones which had been considered of Upper Jurassic or Cretaceous age in nappe 3, regarded by Staub as belonging to the Austrides, were in reality of Lower Carboniferous age, as shown by reef algæ. As the presence of Carboniferous is so far unknown in the Austrides, such a discovery shows that the detailed correlation which has been established for nappes 2 and 3 is premature and ought to be revised on a basis of accurate stratigraphical researches. For this revision I quite agree with Staub that nappe 2 might be correlated with the Liguride Nappe, but only as a working hypothesis.

CONCLUSIONS

It is demonstrated that Corsica belongs to the Alps, the granitic part of this island, with the whole of Sardinia, is a remnant of an Hercynian massif representing the Foreland. The sedimentary region, or Alpine Corsica, shows the *Schistes lustrés* of the zone de Voltri (near Genoa), i.e., metamorphosed sediments deposited in the Alpine geosyncline. Outliers of upper nappes exist, but their correlation with upper nappes of the Alps or with tectonic elements of the Apennines is still a problem to be solved.

CHAPTER IV

SARDINIA [1]

When dealing with the structure of Corsica, we have arrived at the result that the granite massif of Western Corsica, with Sardinia, is part of the Hercynian chain and represents the Foreland of the Alpine geosyncline. As we have recorded effects of the Alpine folding on the Foreland of Corsica we may inquire if the Alpine folding also made itself felt in Sardinia.

Blumenthal has answered this question. Strike-faults are very important in the eastern part of the island, bringing the crystalline substratum, made up of granite or schists, into contact with the sedimentary cover. In certain places a push from the east is demonstrated by the presence of clean-cut thrusts. We have here to deal with *foundation folding*.

The production of this structure was not affected by the vulcanicity, since the latter is much younger in age.

[1] See Fig. 75.

CHAPTER V

THE ISLAND OF ELBA

Termier discovered, in 1910, that the Island of Elba is a region of Alpine nappes. He believed that there were three tectonic elements, as follows (in descending order):

III. A nappe made up of unmetamorphosed sedimentary rocks, with greenstones and radiolarites.

II. A nappe of *Schistes lustrés* with greenstones.

I. A granitic base.

Fig. 80 is a tectonic map, according to Termier, showing these structural elements.

THE GRANITIC BASE

Termier went to Elba after his discovery of the Alpine Nappes in Corsica. There, as already shown, we find a granitic massif of Hercynian age, which can be compared to the Hercynian massifs of the Foreland of the Alps. Termier thought that the granitic massif of Monte Capanne, forming the western part of Elba, played the same rôle and he connected it with the crystalline rocks of the Porto Longone and Calamita region. This was a mistake. The structure of Elba is more complicated owing to the fact that the massif of Monte Capanne is of Tertiary age, the granitic magma having been intruded after the nappes had assumed their position. This massif is equivalent to the one of the Bregaglia (page 195).

In 1931 with Professor Reinhard, of Basle University, we made some geological excursions round the island by boat. On the northern and southern shores of the Monte Capanne Massif beautiful dykes of granite were seen to intrude the radiolarites, limestones and greenstones. The evidence of Tertiary age for this granite is unquestionable. This has also been admitted by Cadisch and Staub. Lotti, the Italian

274

geologist who mapped Elba on the scale of 1 : 25,000, had already shown in 1886 that the granite apophyses are intruded through the sedimentary rocks surrounding the granite massif of Monte Capanne.

The Calamita crystallines, which we saw again in 1932 with Professor Reinhard, are of a different composition. They are gneisses cut by aplite veins bearing tourmaline, and are tilted, showing a dip towards the west. In their upper

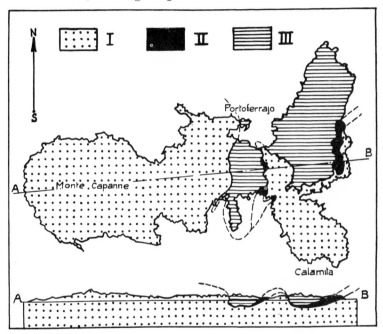

FIG. 80.—Tectonic Map of Elba. *After P. Termier.*
I. Granite Base. II. *Schistes lustrés.* III. Unmetamorphosed Sediments.

part we find a zone of mylonites discovered by Termier. It is very probable that these gneisses belong to a nappe which is the lowest tectonical element of the island. In metamorphosed limestones, belonging to this series of gneisses, occur the well-known iron ores of Calamita.

To summarize : Termier's " granitic base " consists of two entirely different structural elements :

I*a*. The tilted gneisses of the Calamita region, which may belong to a lower nappe.

II*b*. The granitic massif of Monte Capanne of Tertiary age.

The Nappe of *Schistes Lustrés*

There is no doubt that this structural element is to be correlated with a Pennine nappe of the Alps. This nappe consists of :

II*c*. Greenstones.

II*b*. Typical *Schistes lustrés*.

II*a*. White crystalline limestones (Trias), which rest on the mylonites of the Calamita gneisses.

The Upper Nappe

Termier's upper tectonic element shows a very interesting series of sedimentary rocks with ophiolites (serpentine, gabbro, diabase). We find the following succession here, in descending order :

9. Whitish-red limestones, of very fine texture, called Alberese.—Cretaceous.
8. Red and green radiolarites, passing into red and green clays.
7. Ophiolites (gabbros, serpentines and diabases).
6. Whitish-red limestones, like 9.—Cretaceous.
5. Clays with *Posidonomya Bronni.*—Upper Lias.
4. Limestones of very fine texture.—Lower Lias.
3. Dolomitic limestones and breccias.—Trias or Rhaetic.
2. Conglomerates, quartzites, and shales.—Permo-Carbon-iferous.
1. Fossiliferous Paleozoic clays of Silurian age according to Meneghini, and also of Carboniferous age according to Stefani.

Cadisch, in 1929, questioned the upper part (Nos. 6–9) of this series which is very much folded in the sea cliffs to the east of Porto Ferrajo. He thinks that these strata might belong to a separate nappe. I am entirely in agreement with him, since I have seen a crush-breccia of ophiolites and radio-larites at the base of these folded strata. In my opinion there is little doubt that we have to deal here with the rocks of the Liguride Nappe of the Northern Apennines (see page 265). The lower part of the series (Nos. 1–5), with its very peculiar deep-sea facies of the Lower Lias represents a part of the Spezia Nappe of Staub, which frames the window of the Alpi Apuane (page 264).

Crossing, in 1932, with Professor Reinhard from Porto Ferrajo to Golfo Lacona, we were fortunate in finding a series of strata which represents a normal sequence, as follows : 1, red radiolarites and clays of a deep-sea facies, of Jurassic age. 2, " Alberese " or whitish-red limestones also of a deep-sea facies. The foraminifera show a Lower Cretaceous age. This formation passes gradually into : 3, black limestones, of a grey hue, of a deep-sea facies, containing *Actiniscus* characteristic of the Upper Cretaceous. Here the ophiolites cap the sedimentaries. This series may correspond to that of the Liguride Nappe of the Apennines though Upper Cretaceous has not yet been found in the latter.

CONCLUSIONS

The Island of Elba is a country of Alpine nappes, as demonstrated by Termier, but its structure is more complicated than he thought. Instead of two nappes there may be four, from top to bottom :

4. A nappe of ophiolites and radiolarites, equivalent to the Liguride Nappe of the Northern Apennines.

3. A nappe which is the homologue of the Spezia Nappe of the window of the Alpi Apuane.

2. A nappe of *Schistes lustrés*, demonstrating the presence of rocks deposited in the Alpine geosyncline.

1. The gneisses of Calamita which may represent one of the Pennine Nappes of the Alps.

This structure being the result arrived at during reconnaissance work must only be accepted as a *working hypothesis*. Though Lotti's mapping of 1886 was a splendid piece of exploration a revision based on modern principles is now of paramount importance to get a good idea of the tectonics.

CHAPTER VI

THE ALPINE CHAIN OF SOUTHERN SPAIN

INTRODUCTION

In dealing with the structure of the Alpine Chain of Southern Spain we need a general introduction to the geology of Spain and Portugal.

The main geological features are :

1. The Meseta.
2. The Alpine belt.

The Meseta is made up of the oldest structural elements of Spain and Portugal, in which can be recognized a Pre-Cambrian region, forming the north-western part of the Iberic Peninsula. Caledonian folding is represented on its eastern border, but we do not know much about it. An Hercynian belt frames the Pre-Cambrian block to the east and south, thus forming an arc round the Sierra de Guadarrama.

The southern and eastern parts of the Meseta show structures which according to R. Staub are only a **reaction** of the Alpine folding, whilst Argand sees in them foundation folding (see page 17). If the difference between these views is great, their authors nevertheless, agree in considering the Meseta as the Foreland of the Alpine Chain, which latter is represented by the mountains of Southern Spain (Betic and Sub-Betic chains) in the provinces of Andalusia, Granada and Murcia. Thus the Pyrenees, the Cantabrian Mountains, Sierra de la Demanda, Montes Universales and Sierra Morena belong to the Foreland of the Alps. It is worth noting that *the Pyrenees have not been formed in the Alpine geosyncline*, contrary to the opinion of several authors. Indeed, one misses in the Pyrenees all the characteristic rocks of the Alpine geosyncline. The sedimentary rocks of the Pyrenees are only the sedimentary cover of the Meseta. Moreover, the structure of the Pyrenees belongs to the type of structure of the Foreland and not to the great recumbent folds of the Alpine geosyncline.

The depression of the river Guadalquivir, which has been described by several authors as a "graben," separating the horst of the Meseta (Sierra Morena) from the horst of the Betic Cordillera, must be considered rather as the frontal plain of the Alpine chain. Indeed, Staub compares it with the Swiss plateau at the front of the Alps.

The Alpine Belt. To the south of the Meseta, viz., to the south of a line stretching from Huelva to Cape Nao and along the river Guadalquivir, we find a great mountainous region 600 kilometres long, 120–160 kilometres wide, with peaks 9,000 feet high, which forms the border of the Mediterranean Sea, from Gibraltar to Cape Nao. This is the Alpine belt, which can be divided into two parts:

A. *The Betic Chain* along the sea, made up of crystalline rocks. It extends from Gibraltar to Cartagena and includes the *Sierra Nevada*.

B. *The Sub-Betic Chain* to the north-west of the former, made up of sedimentary rocks.

The first stratigraphical studies in the Alpine belt of Spain are due to Spanish geolo-

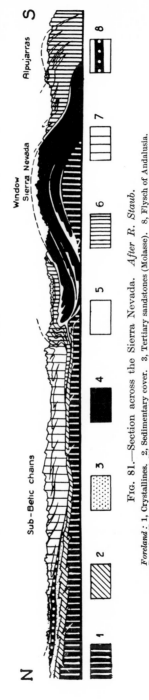

Fig. 81.—Section across the Sierra Nevada. *After R. Staub.*

Foreland: 1, Crystallines. 2, Sedimentary cover. 3, Tertiary sandstones (Molasse). 8, Flysch of Andalusia.
Pennine Nappes: 4, Crystallines. 5, Sedimentary.
Austrides: 6, Alpujarrides.
7, Sub-Betic (Trias-Cretaceous) of the Betic Nappe of Malaga.

gists, as Mallada, Cortazar, and Gonzalo y Tarin. To Nickles and Douvillé, two French geologists, we owe our first knowledge of the tectonics of the Sub-Betic Chain, based on detailed stratigraphical researches. Suess, Termier, Kober, Staub and Stille correlated the Betic and Sub-Betic chains with the Alps (Fig. 81).

THE ALPINE CHAIN OF SPAIN

The authors who have been dealing with the structure of the Alpine Chain of Spain are R. Douvillé, Brouwer and his pupils, R. Staub, Klebelsberg, Blumenthal and Fallot; quite an international gathering of geologists.

In 1926, R. Staub in his paper, *Gedanken zur Tektonik Spaniens*, published a general view on the tectonics of the Alpine Chain of Spain, which must be considered as a very interesting pioneer work. Indeed, one is strongly impressed in reading Staub's conclusions, but if one tries to check Staub's hypothesis with the facts discovered lately after accurate mapping by Brouwer and his pupils, Blumenthal and Fallot, one finds that the picture produced by Staub of the Alpine Chain of Spain as shown in Fig. 81 is too simple. On the ground-work of Staub's working hypothesis, Blumenthal published new ideas on the relations between the crystalline (Betic belt) and the sedimentary part (Sub-Betic belt) of the Alpine Chain.

Let us now try to give a summary of the results recently produced.

THE BETIC CHAIN

Again, as in the Alpi Apuane a *window* plays a very great rôle in the study of the Betic Chain. It is the *Sierra Nevada window*, due to erosion on an axis culmination. This window has been compared also, with reason, to the *Tauern window* in the Eastern Alps (p. 32). Termier, in 1911, had foreseen this great result.

Brouwer and his pupils arrived at the conclusion that the *Sierra Nevada* appears in the window as a great carapace made up of schists. The frame of the window is represented by three nappes of schists separated from one another by Alpine Trias. These nappes override the Sierra Nevada and, at the

contact between the latter and the lowest nappe, occurs a
" crushzone," viz., a zone in which schists and metamorphosed
mesozoic rocks are mixed owing to crushing.   An upper nappe
(No.4) has been later discovered by Blumenthal, the names
given are, as follows, from top to bottom :

4. Betic of Malaga.
3. Guajar ⎫
2. Lanjaron ⎬ Alpujarrides.
1. Lujar ⎭

As nappes 1–3 **root** in the region of las Alpujarras, they
are named " *Alpujarrides.*"

Owing to the facies of their Trias, these nappes are said to
represent the Austrides.   Hence the carapace of the Sierra
Nevada corresponds to the Pennine Nappes, for the Mesozoic of
the crushzone, is, according to Brouwer, of a Pennine facies.

To the south-west the Betic of Malaga has been intruded
by a basic magma which is well known under the name of
Peridotites of the Serrania de Ronda.   According to Blumenthal,
the basic magma digested the rocks of the nappes, while these
were in process of formation.

The age of the intrusion is Pre-Middle Eocene for the Middle
Eocene (Flysch) was not yet deposited.

The Peni-Betic Belt.   Whereas the nappes of the Alpujarras do not show any unmetamorphosed sedimentary rocks
other than the Trias, the crystallines of the Betic of Malaga
support : Trias, Jurassic, Cretaceous and Lower Tertiary.
These rocks form a belt on the frontal part of the nappe of the
Betic of Malaga, which has been named by Blumenthal : *the
Peni-Betic Belt.*

THE SUB-BETIC AND PRE-BETIC BELTS

The Peni-Betic Belt overrides to the north a sedimentary
complex made up of rocks stratigraphically extending from
the Trias to the Tertiary.   Fig. 82 shows Blumenthal's interpretation, according to which the Pre-Betic and Sub-Betic
Belts are the sedimentary cover of the Foreland, which has
been peeled off and dragged northwards owing to the travelling
of the Betic and Peni-Betic Nappes.   Thus the Pre-Betic and
Sub-Betic Nappes might be compared, to some extent, with the
High Calcareous Alps of Switzerland (see page 58).   R. Staub's

explanation of 1926 was much simpler than the above. He regarded the great sedimentary belt to the north of the Sierra Nevada as representing only the sedimentary of the Betic Nappe of Malaga. It is on a basis of stratigraphical studies that Blumenthal has put forward the new explanation which we have summarized.

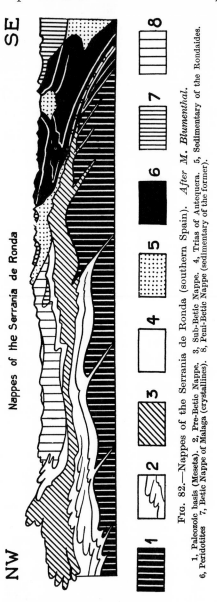

FIG. 82.—Nappes of the Serrania de Ronda (southern Spain). *After M. Blumenthal.*

1, Paleozoic basis (Meseta). 2, Pre-Betic Nappe. 3, Sub-Betic Nappe. 4, Trias of Antequera. 5, Sedimentary of the Rondaides. 6, Peridotites 7, Betic Nappe of Malaga (crystallines). 8, Peni-Betic Nappe (sedimentary of the former).

SE

NW

Nappes of the Serrania de Ronda

THE AGE OF FOLDING IN THE ALPINE CHAIN OF SPAIN

According to Blumenthal (1934), the folding in the Alpine Chain of Spain took place as follows :

1. A first phase of folding of Pre-Eocene age (very likely Upper Cretaceous).

2. The transgression of the sea of the Flysch with very thick deposits of Lutetian to Middle Oligocene age.

3. The paroxysm of folding, of Upper Oligocene age, to which is due the travelling northwards of the nappes of the Alpujarrides, of the Betic of Malaga, and also the laminating and dragging of the nappes of the Peni and Sub-Betic Belts.

4. A phase of erosion followed by the transgression of the Lower Miocene (Burdigalian).

5. A phase of emersion followed by little folding (posthumous).

6. The Pliocene transgression and Post-Pliocene u.heaval.

THE RELATIONS BETWEEN THE ALPINE CHAIN OF SPAIN AND THE RIF IN MOROCCO

Argand suggested that the Alpine Chain of Spain unites with the Rif in Morocco, following the morphological bending of Gibraltar.   According to Termier and Staub, the Alpine Chain of Spain strikes east-west and terminates in the Atlantic.

Marin, Blumenthal and Fallot have lately pointed out that the Paleozoic and Mesozoic of the northern part of Morocco (Rif) are the exact homologue of the formations of the same age of the Betic Nappe of Malaga.   Moreover, these authors have recorded that Peni-Betic elements of Andalusia occur at Gibraltar and to the west of Algeciras.   If the Alpine Nappes of Andalusia travelled towards the Foreland (north), in Morocco there is evidence of movements towards the west, the south-west and the south, but on a much smaller scale than in Andalusia. Whilst the results arrived at by Marin, Blumenthal and Fallot must at present only be considered as a very interesting *working hypothesis*, they do seem to agree with Argand's suggestion.

CHAPTER VII

THE BALEARIC ISLANDS

INTRODUCTION

The bathymetrical map of the western Mediterranean Sea shows that the Balearic Islands are connected to Spain by a submarine ridge, 250 kilometres long. The Island of Iviza rises about 475 metres above sea-level, Majorca, 1,443 metres and Minorca 358 metres.

Geographers generally consider Majorca and Minorca as forming a unit. Tectonically Majorca is connected to Iviza and not to Minorca. This latter island, as pointed out by Suess, is entirely different from a structural point of view.

Majorca is the greatest of the Balearic Islands, and is also geologically the more important owing to the structures shown in the Sierra Norte, 90 kilometres long, with summits rising 1,000–1,400 metres directly above sea-level.

MAJORCA

The Island of Majorca can be divided into three parts :

1. The Sierra Norte, forming the northern part.
2. An intervening plain.
3. The Sierra de Levante, in the south.

The northern coast of Majorca presents some of the finest scenery of the western Mediterranean region. The Sierra terminates in an impressive scarp which plunges abruptly into the sea. This scarp is divided into two cliffs by a horizontal ledge of some width, which the road follows and on which the cultivation is concentrated.

The lower cliff, 1,000–1,500 feet high, is generally made up of Lower Trias and is often of a red colour owing to the presence of Buntsandstein.

The ledge is composed of Tertiary which stratigraphically overlies the Trias and, in some cases, the Jurassic.

In 1910, Collet showed that the upper cliff of Trias and Jurassic had been thrust over the Tertiary of the ledge. He thus demonstrated that the northern part of the Sierra consisted of two tectonic elements. This made a starting-point for the studies on the structure of the whole island, carried out chiefly by Fallot in the Sierra Norte and by Darder in the Sierra de Levante. Fallot's valuable memoir and geological

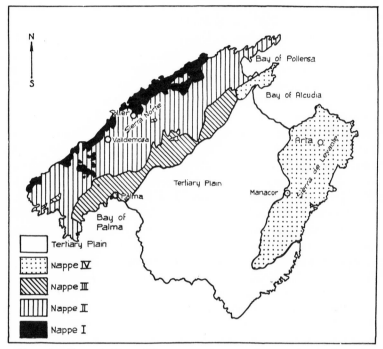

FIG. 83.—Tectonic Map of Majorca. *After P. Fallot and Darder Pericas.*

map have become indispensable to the worker in the Balearic Islands.

Fallot's work has shown that Majorca is made up of four nappes. The three lower elements (I–III) are confined to the Sierra Norte and the fourth (IV) forms the Sierra de Levante. The latter nappe has very likely been thrust over the upper element of the Sierra Norte as shown by an outlier of nappe IV resting on nappe III in the north-eastern part of the Sierra.

Moreover, Fallot has discovered the existence of three nappes in Iviza. We will have to correlate the structures found in the Balearic Islands and to study what they represent in the Alpine Chain.

In general, we can say that the tectonics are characterized by clean-cut thrusts and not by recumbent folds.

The nappes. Nappe II, which has been thrust over nappe I, is the most important for the whole mountainous part of the Sierra Norte has been eroded out of it. Little would be known of nappe I, which crops out in the sea cliff along the northern coast, if Fallot had not discovered windows in nappe II, cut on an axis culmination, which reveal that the substratum to the latter nappe is formed by nappe I. The contact between nappe II and nappe I may be marked either by a simple mechanical contact of Trias overriding the Tertiary, or by slices which were dragged along the thrust-plane and intercalated between the two nappes.

Nappe I is made up of : 1, Lower Trias ; 2, Rare outcrops of reduced Jurassic ; 3, Tertiary generally covered by cultivation.

Nappe II consists of : 1, An important series of middle and Upper Trias ; 2, Jurassic limestones forming the main ridge ; 3, Fossiliferous Lower and Middle Cretaceous ; 4, Middle and Upper Oligocene (black conglomerates). The Cretaceous and Oligocene have been only preserved in rare synclines.

Fig. 84 shows the relations between nappe II and nappe I in a window. This section may be considered as an exception, for nappe II forms generally a great carapace, little folded, covering nappe I. In Fig. 84 we see imbrications of nappe II. This phenomenon occurs in the main ridge in the region of Valldemosa and may be followed as far as Soller. This structure is very likely due to the fact that the travelling of nappe II was affected by an obstacle.

The southern part of the Sierra Norte, which dominates the Bay of Palma, is formed by nappe III, as seen in Fig. 84. According to Fallot's profiles, the folds of nappe II are cut by the thrust-plane of nappe III.

Nappe IV is limited to the Sierra de Levante. Darder, who discovered this tectonic element, is inclined to see several nappes where Fallot admits only one. It is interesting to remark that nappe IV shows imbrications in the region of Arta which seem to correspond to the region of the Sierra Norte, where imbrications also occur.

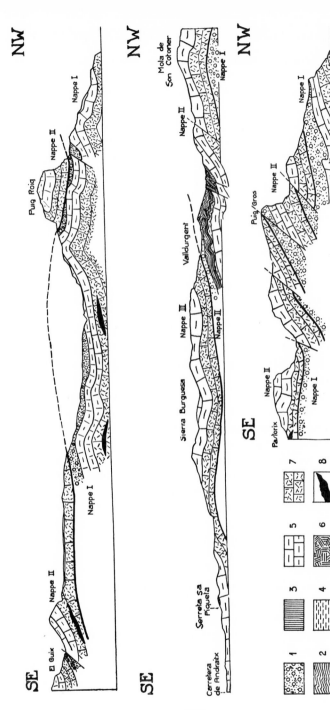

FIG. 84.—Section across the Cordillera Norte of Majorca. *After P. Fallot.*

1, Tertiary. 2, Neocomian. 3, Upper Jurassic. 4, Upper Lias. 5, Middle Lias. 6, Keuper. marls. 7, Middle and Upper Trias, dolomitic limestones. 8, Eruptive rocks.

The Island of Cabrera, to the south-west of Majorca, according to Fallot, belongs to nappe IV.   Also the region to the
north of the Bay of Alcudia (Fig. 83) may represent an outlier
of nappe IV.

From a stratigraphical point of view nappe IV is different
from the nappes of the Sierra Norte, as the former shows facies
of a more southern character.

The movement of all the nappes of Majorca is directed from
south-east to north-west.   The development of all the nappes
of Majorca is of Post-Burdigalian age.

IVIZA

According to Fallot, Iviza is equivalent to Majorca.   Indeed,
three nappes have been recorded in the former as in the Sierra
Norte of Majorca.   *Geologically Iviza and Majorca form a unit.*

MINORCA

Though Minorca is not a country of nappes, the study of its
tectonics, as shown by Fallot, proves three kinds of movements :

1. An Hercynian movement demonstrated by the folding of
the Devonian.

2. A Pre-Burdigalian movement exemplified by folds and
the overriding of the Trias by the Devonian.

3. A Post-Burdigalian movement which produced only a
tilting of the Burdigalian.

From a stratigraphical point of view, Minorca differs from
Majorca and Iviza owing to the presence of Devonian, partly
of continental facies.   Furthermore, from the end of the Trias
onwards the paleogeography of Minorca is entirely different
from that of Majorca and Iviza.   Thus Minorca must be
separated from the other Balearic Islands.

CONCLUSIONS ON THE BALEARIC ISLANDS

In Iviza and Majorca we are dealing with the remnant of the
extremity of the Sub-Betic Chain of Southern Spain.   Our
knowledge of the structure of the latter being far from complete, a correlation with the Balearic Islands leads us to some
speculation.   As a working hypothesis, Fallot expressed the
view that the lower nappe of Iviza might be connected with
the region of Spain situated between Alicante and Denia.
Moreover, the Sierra Norte of Majorca might be considered as

the homologue of the imbrications of the Sierra of Caravaca. The Sierra de Levante of Majorca is either a more southern part of the nappes of the Sierra Norte or a well-defined upper element of which the roots now lie under the sea, to the south of Majorca. *Minorca does not belong to the Sub-Betic Chain of Spain.* It is either Foreland or Hinterland. The bending of the submarine ridge between Minorca and Majorca seems to show that the Sub-Betic Chain, represented by Majorca, passes to the south-west of Minorca. If that is the case, Minorca belongs to the Foreland.

BIBLIOGRAPHY

1. LOTTI, B.—Carta geologica dell'Isola d'Elba, 1 : 25,000. R. Ufficio Geologico, Roma, 1884.
2. SACCO, F.—L'Apennino settentrionale. Boll. Com. geol. Ital. Torino, 1904.
3. STEINMANN, G.—Alpen und Apennin. Zeitschr. Deutsche geol. Ges., 1907.
4. TERMIER, P.—Sur les relations de l'île d'Elbe, des Alpes et des Dinarides. Bull. Soc. géol. France, 1907.
5. TERMIER, P.—Sur la tectonique de l'île d'Elbe. Bull. Soc. géol. France, 1910, p. 134.
6. LOTTI, B.—Geologia della Toscana. Mem. desc. Carta geol. Italia, 1910.
7. TILMANN, N.—Zur Tektonik des Südapennin. Geol. Rundschau, 1912.
8. STEINMANN, G.—Über Tiefenabsätze des Oberjura im Apennin. Geol. Rundschau, 1913, pp. 572–576.
9. FALLOT, P.—Etude géologique de la Sierra de Majorque. Ch. Béranger, Paris et Liège, 1922.
10. FALLOT, P.—Le problème de l'île de Minorque. Bull. Soc. géol. France, T. XXIII, 1923.
11. FUCINI, A.—Studii geologici sul Monte Pisano. Catania, 1924–5.
12. BROUWER, H. A.—Zur Tektonik der Betischen Kordilleren. Geol. Rundschau, XVII, 1926, p. 332.
13. BROUWER, H. A.—Zur Geologie der Sierra Nevada, Ibid., 1926, p. 118.
14. STEINMANN, G.—Gibt es fossile Tiefseeablagerungen von erdgeschichtlicher Bedeutung ? Ibid., 1925, p. 435.
15. DARDER, PERICAS.—La Tectonique de la région orientale de l'île de Majorque. Bull. Soc. géol. France, Paris, 1925.
16. STAUB, R.—Gedanken zur Tektonik Spaniens, Vierteljahrsschrift Naturf. Ges. Zurich, 1926, pp. 196–261.
17. TILMANN, N.—Tektonische Studien in der Catena metallifera Toscanas. Geol. Rundschau. Steinmann-Festschrift, 1926.
18. STILLE, H.—Über westmediterrane Gebirgszusammenhänge. Abh. Ges. Wiss. Göttingen, math. Kl., N.F. XII, 3, 1927.

19. STEINMANN, G.—Die ophiolithischen Zonen in den mediterranen Kettengebirgen. C. R. Congr. géol. intern., Madrid, 1927.

20. STAUB, R.—Der Bewegungsmechanismus der Erde. Borntraeger, Berlin, 1928.

21. STAUB, R.—Der Deckenbau Korsikas und sein Zusammenhang mit Alpen und Apennin. Vierteljahrsschr. d. Naturf. Gesell. Zürich, 1928, pp. 298–348.

22. TERMIER, P. et ∫ Nouvelles observations géologiques dans la Corse
    MAURY, E.   ⌊ orientale. C. R. Acad. Sc., t. 186, Paris, 1928.

23. CADISCH, J.—Zur Geologie der Insel Elba. Verh. Naturf. Ges. Basel, Bd. XL, 1929, pp. 52–61.

24. MARIN, A.,   ⌉ Observations géologiques sur le nord-ouest
    BLUMENTHAL, M. et ⌋ du Rif marocain. Bull. Soc. géol. France,
    FALLOT, P.   t. XXX, 1930.

25. FALLOT, P.—Etat de nos connaissances sur la structure des chaines Bétique et Subbétique (Espagne méridionale). Livre Jubilaire Cent. Soc. géol. France, 1930, pp. 281–305.

26. SEIDLITZ, V. W.—Diskordanz und Orogenese der Gebirge am Mittelmeer. Borntraeger, Berlin, 1931.

27. KOBER, L.—Das alpine Europa. Borntraeger, Berlin, 1931.

28. STAUB, R.—Die Stellung Siziliens im mediterranen Gebirgssystem. Vierteljahrsschrift Naturf. Ges. Zurich, LXXVII, 1932, pp. 159–182.

29. ZACCAGNA, DOMENICO.—Descrizione Geologica delle Alpi Apuane. Mem. descritt. Carta Geol. Italia, XXV, Roma, 1932.

30. STAUB, R.—Die Bedeutung der Apuanischen Alpen im Gebirgsbau der Toskana nebst einigen Gedanken zur Tektonik des Apennins. Vierteljahrsschrift d. Naturf. Ges. Zurich, LXXVII, 1932.

31. BLUMENTHAL, MORITZ M.—Geologie der Berge um Ronda (Andalusien). Eclogae geol. Helvet., 1933, vol. XXVI, Nº 1, pp. 43–92.

32. BLUMENTHAL, MORITZ M.—Die Grenzverhältnisse zwischen sub- und penibetischer Zone im Grenzgebiet der Provinzen Malaga, Sevilla und Cadiz (Strecke Almargen-Olvera). Eclogae geol. Helvet., 1934, vol. XXVII, Nº 1, pp. 147–180.

33. MERLA, G.—I granite della formazione ofiolitica apenninica. Bolletino R. Ufficio geologico d' Italia. Vol. 58, Roma, 1934.

INDEX

Aar; Granite, Central zone of, 49, **51**, 99, 109; massif, 4, 5, 31, 34, **49**, 63, 64, 81, 82, 87, 97, 99, 100, 109, 153, 154, 166, 185, 199; valley, 81, 82, 87, 89.
Adelboden, 87, 92, 239, 240.
ADRIAN, H., 86, 105.
Adriatic Subsidence, Phase of, 22.
Adula; gneiss, 158, 160; nappe, 145.
Aermighorn, 87.
Africa, 25, 27.
Agassizjoch, 51.
Aiguilles Rouges de Chamonix, 4, 17, 31, 34, **36**, 59, 60, 62, 65, 75, 78, 81, 84, 97, 127; Carboniferous of the, 40; granite, 40; The Rock complexes, **38, 39**; Trias, of the, 79.
Airolo, 54.
Aix les Bains, 115.
Alagna valley, 171.
Alberese, 265, 277.
Albristhorn, 239.
Albula; digitation, 210, 212; granite, 210–212; tunnel, 211.
Aletsch glacier, 52.
Aletschhorn, 52, 75.
Algeria, 25.
Allalinhorn, 175, 187; Klein, 175.
Allée, Col de l', 184.
Allgaü facies, 226; Nappe, 225.
Alphubeljoch, 186.
Alpides, 22, 23.
Alpiglen, 113.
Alps, Austrian, 26; Cottian, 1; Dinaric, 1; Gailtaler, 4; Graian, 4; High Calcareous, 3–5, 10, 12, 16, 19, 22, 27, 31–34, 44, **58**, 109, 233, 234, 238, 242, 251, 281; Lepontine, 4; Ligurian, 4; of Sixt, 78; Pennine, 4, 10, 20, 26, 54, 198, 203, 229, 238; roots of, 203; roots of the nappes of the High Calcareous, 99; Southern Limestones, 4; Western, 145.
Alpujarrides, 281.
Alte Caserne, 157.
Altels, 81.

Alv, Piz, 214, 219.
Ambin, 148, 175, 176.
Ammertengrat, 96.
AMPFERER, O, 230.
AMSLER, A, 142, 143.
Andalusia, 1.
ANGEL, F., 231.
Ankenballi, 50.
Ankérite, 61.
Ankogel Nappe, 177, 183, 184.
Annecy, Lake of, 116, 248.
Annemasse, 250.
Annes, the, 248.
Anniviers valley, 171, 173, 199.
Anterne, Col d', 60, 78.
Anticline, overfolded, 9, 135; recumbent, 9, 10; symmetrical, 8, 135; trunk, 134; unsymmetrical, 8, 135.
Antigorio gneiss, 157, 160; nappe, 145, 154, **155**, 156, 203.
Anzeindaz, 84.
Aosta valley, 145, 183, 185, **200**.
Apennines, 1, 19, 22, 25, 256, 258, 260, 265–268, 271, 272.
Apet, Col de l', 180.
Apuane, Alpi, 25, 258, **260**, 261, 262, 265, 267, 268, 276.
Aravis, 4.
Arbedo, 203.
Arbenhorn, 240.
ARBENZ, P., 7, 12, 28, 58, 64, 89, 91, 96, 103, 105, 110, 113, 206.
Arblatsch, 184.
Ardenno, 224.
Ardez, 217, 218.
Ardon, 84.
Arezzo, 264, 266, 268.
ARGAND, E., 1, 5–7, 13, 17, **19**, 20–28, 58, 137, 145, 148, 149, 152, 160, 165, 166, 169, 171, 172, 176, 178–180, 184, 185, 187, 188, 190–192, 199, 200, 203–207, 241, 257, 258, 261, 271, 278, 283.
Argentera, *see* Mercantour.
Argentière, Aiguille d', 37.
Arlas Munt, 213.
ARNI, P., 230.

291